Peter Berck   Knut Sydsæter

# Economists' Mathematical Manual

Second Edition

Springer-Verlag

Berlin Heidelberg New York
London Paris Tokyo
Hong Kong Barcelona
Budapest

Professor Dr. PETER BERCK
Department of Agricultural and Resource Economics
University of California, Berkeley
207 Giannini Hall
Berkeley, CA 94720, USA

Professor KNUT SYDSÆTER
Department of Economics
University of Oslo
P.O.B. 1095, Blindern
N-0317 Oslo, Norway

ISBN 3-540-56374-1 Springer-Verlag Berlin Heidelberg New York Tokyo
ISBN 0-387-56374-1 Springer-Verlag New York Heidelberg Berlin Tokyo

ISBN 3-540-54370-8 1. Auflage Springer-Verlag Berlin Heidelberg New York Tokyo
ISBN 0-387-54370-8 1st Edition Springer-Verlag New York Heidelberg Berlin Tokyo

Library of Congress Cataloging-in-Publication Data. Berck, Peter, 1950 – Economists' mathematical manual / Peter Berck, Knut Sydsæter. -- 2nd ed. p.cm. Includes bibliographical references and index.
ISBN 0-387-56374-1 (U.S. : alk. paper)
1. Economics, Mathematical. I. Sydsæter, Knut. II. Title.
HB135.B467 1993 330'.01'51--dc 20 92-46065 CIP

This work is subject to copyright. All rights are reserved, whether the whole or part of the material is concerned, specifically the rights of translation, reprinting, reuse of illustrations, recitation, broadcasting, reproduction on microfilms or in other ways, and storage in data banks. Duplication of this publication or parts thereof is only permitted under the provisions of the German Copyright Law of September 9, 1965, in its version of June 24, 1985, and a copyright fee must always be paid. Violations fall under the prosecution act of the German Copyright Law.

© Springer-Verlag Berlin · Heidelberg 1991, 1993
Printed in Germany

The use of registered names, trademarks, etc. in this publication does not imply, even in the absence of a specific statement, that such names are exempt from the relevant protective laws and regulations and therefore free for general use.

Printing: Betz-Druck, Darmstadt
Bookbinding: J. Schäffer GmbH u. Co. KG., Grünstadt
2142/7130-543210 – Printed on acid-free paper

# Preface

The practice of economics requires a wide-ranging knowledge of formulas from mathematics and mathematical economics. The selection of results from mathematics included in handbooks for chemistry and physics ill suits economists. There is no concise reporting of results in economics. With this volume, we hope to present a formulary, targeted to the needs of students as well as the working economist. It grew out of a collection of mathematical formulas for economists originally made by Professor B. Thalberg and used for many years by Scandinavian students and economists.

The formulary has 32 chapters, covering calculus and other often used mathematics; programming and optimization theory; economic theory of the consumer and the firm; risk, finance, and growth theory; non-cooperative game theory; and elementary statistical theory.

The book contains just the formulas and the minimum commentary needed to re-learn the mathematics involved. We have endeavored to state theorems at the level of generality economists might find useful. By and large, we state results for $n$-dimensional Euclidean space, even when the results are more generally true. In contrast to the economic maxim, "everything is twice more continuously differentiable than it needs to be", we have listed the regularity conditions for theorems to be true. We hope that we have achieved a level of explication that is accurate and useful without being pedantic.

In a reference book, errors are particularly destructive. We would like to thank E. Biørn, P. Frenger, G. Judge, D. Lund, M. Machina, H. Mehlum, K. Moene, G. Nordén, T. Schweder, A. Seierstad, L. Simon, and B. Øksendal for useful suggestions and for their help in locating our mistakes. We hope that readers who find our remaining errors will call them to our attention so that we may purge them from further editions. We are, of course, solely responsible for whatever errors remain.

Professor Arne Strøm has written the TeX macros for this book and offered expert help both technically and mathematically.

Berkeley and Oslo in June 1991

*Peter Berck,   Knut Sydsæter*

## Preface to the second edition

In this edition we have made a number of corrections and some minor additions. We thank Professor Peter Hammond for pointing out some rather embarrassing errors and cand. oecon. Halvor Mehlum for very valuable suggestions.

Berkeley and Oslo in August 1992

*Peter Berck, Knut Sydsæter*

# Contents

1. **Functions of one variable. Complex numbers** .................... 1
   Roots of quadratic and cubic equations. Polynomials. Descartes's rule of signs. Classification of conics. Asymptotes. Newton's approximation method. Powers, exponentials, and logarithms. Trigonometric and hyperbolic functions. Complex numbers.

2. **Limits. Differentiation (one variable)** ............................ 9
   Limits. Continuity. The intermediate value theorem. Differentiable functions. General and special rules. Mean value theorems. L'Hôpital's rule. Differentials.

3. **Partial derivatives** ................................................ 13
   Partial derivatives. Chain rules. Differentials. Slopes of level curves. Implicit function theorem. Homogeneous and homothetic functions. Gradients and directional derivatives.

4. **Elasticities. Elasticities of substitution** ......................... 17
   General and special rules. Directional elasticities. Marginal rate of substitution. Elasticity of substitution for functions of two and $n$ variables. Allen-Uzawa's elasticity of substitution. Morishima's elasticity of substitution.

5. **Systems of equations** .............................................. 21
   General systems of equations. Jacobians. General implicit function theorem. Degrees of freedom. The counting rule. Functional dependence. Local and global inverse function theorems. Gale-Nikaido theorems. Contraction mapping theorem. Brouwer and Kakutani's fixed point theorems. General results on linear systems of equations.

6. **Inequalities** ..................................................... 27
   Inequalities for arithmetic, geometric, and harmonic means. Inequalities of Hölder, Cauchy-Schwarz, Chebychev, Minkowski, and Jensen.

7. **Series. Taylor formulas** .................................................. 31

   Arithmetic and geometric series. Convergence criteria. Maclaurin and Taylor formulas. Binomial coefficients. Newton's binomial formula. Summation formulas.

8. **Integration** ..................................................................... 35

   General and special rules. Convergence of integrals. Comparison test. Leibniz's formula. The Gamma function. Stirling's formula. The Beta function. Trapezoid formula. Simpson's formula. Multiple integrals.

9. **Difference equations. (Recurrence relations.)** ..................... 43

   Solutions of linear equations of first, second, and higher order. Stability. Schur's theorem. Matrix formulations.

10. **Differential equations** .................................................... 47

    Separable, projective, and logistic equations. Linear first-order equations. Bernoulli and Riccati equations. General linear equations. Variation of parameters. Stability for linear equations. Routh-Hurwitz's criterion. Autonomous systems. Local and global stability. Liapunov theorems. Lotka-Volterra models. Local saddle point theorem. Local and global existence theorems. Partial differential equations of the first order.

11. **Topological concepts in Euclidean space** ........................... 55

    Point set topology. Convergence of sequences. Cauchy sequences. Continuous functions. Sequences of functions. Correspondences. Lower and upper hemicontinuity. Infimum and supremum.

12. **Convexity** ..................................................................... 59

    Convex sets. Separation theorems. Concave and convex functions. Hessian matrices. Quasi-concave and quasi-convex functions.

13. **Classical optimization** .................................................... 65

    Basic definitions. First-order conditions. Second-order conditions. Constrained optimization. Lagrange's method. Properties of Lagrange multipliers.

14. **Linear and nonlinear programming** ................................... 71

    Basic results in linear programming. Duality. Shadow prices. Complementary slackness. Kuhn-Tucker theorems. Saddle point results. Quasi-concave programming. Properties of value functions. Nonnegativity conditions.

15. **Calculus of variations and optimal control theory** ................ 77

    The simplest problem. Euler's equation. Legendre's condition. Transversality conditions. Sufficient conditions. Scrap value functions. More general variational problems. Control problems with fixed time interval. The maximum principle.

Mangasarian and Arrow's sufficiency conditions. Properties of the value function. Free terminal time problems. Scrap value functions. Current value formulations. Linear quadratic problems. Infinite horizon problems.

**16. Discrete dynamic optimization** .................................. 85

Dynamic programming. The fundamental equation. Infinite horizon. Discrete optimal control theory.

**17. Vectors. Linear dependence. Scalar products** ...................... 89

Linear dependence and independence. Subspaces. Bases. Scalar products. Norm of vectors. Angles between vectors in $R^n$.

**18. Determinants** ...................................................... 93

$2 \times 2$- and $3 \times 3$-determinants. General determinants and their properties. Cofactors. Vandermonde and other special determinants. Minors. Cramer's rule.

**19. Matrices** .......................................................... 97

Special matrices. Matrix operations. Inverse matrices and their properties. Trace. Rank. Matrix norms. Partitioned matrices.

**20. Characteristic roots. Quadratic forms** ............................ 103

Characteristic roots and vectors. Diagonalization. Spectral theory. Jordan decomposition. Cayley-Hamilton's theorem. Quadratic forms and criteria for their signs. Singular value decomposition. Simultaneous diagonalization.

**21. Special matrices. Leontief systems** .............................. 109

Properties of idempotent, orthogonal, and permutation matrices. Nonnegative matrices. Frobenius roots. Decomposable matrices. Leontief systems. Hawkins-Simon conditions. Dominant diagonal matrices.

**22. Kronecker products and the vec operator** ......................... 113

Definition and properties of Kronecker products. The vec operator and its properties.

**23. Differentiation of vectors and matrices** ......................... 117

Differentiation of vectors, matrices, and determinants with respect to elements, vectors, and matrices.

**24. Comparative statics. Value functions** ............................ 119

Equilibrium conditions. Comparative statics. The value function. Envelope results.

## 25. Properties of cost and profit functions .......... 123
Cost functions. Conditional factor demand functions. Shephard's lemma. Profit functions. Factor demand functions. Supply functions. Hotelling's lemma. Puu's equation. Special functional forms and their properties. Cobb-Douglas, CES, Law of the minimum, Translog cost functions.

## 26. Consumer theory .......... 129
Utility maximization. Indirect utility functions. Consumer demand functions. Roy's identity. Expenditure functions. Hicksian demand functions. Slutsky equation. Equivalent and compensating variations. Special functional forms and their properties. AIDS, LES, and Translog indirect utility function.

## 27. Topics from finance and growth theory .......... 135
Compound interest. Effective rate of interest. Present value calculations. Internal rate of return. Norstrøm's rule. Solow's growth model. Ramsey-type problem.

## 28. Risk and risk aversion theory .......... 139
Absolute and relative risk aversion. Arrow-Pratt risk premium. Stochastic dominance of first and second degree. Hadar and Russell's theorem. Rothschild-Stiglitz's theorem.

## 29. Finance and stochastic calculus .......... 141
Capital asset pricing model. Black and Schole's option pricing model. European call option. Stochastic integrals. Ito's formulas. A stochastic control problem. Hamilton-Jacobi-Bellman's equation.

## 30. Non-cooperative game theory .......... 145
An $n$-person game. Nash equilibrium. Mixed strategy extension of a game. Two-person games. Minimax theorem. Exchangeability property.

## 31. Statistical concepts .......... 149
Probability. Bayes rule. Expectation. Variance. Covariance. Correlation coefficient. Chebychev's inequality. Estimators. Bias. Marginal and conditional densities. Testing. Power of a test. Type I and type II errors. $\alpha$-level of significance.

## 32. Statistical distributions. Least squares .......... 153
Binomial, multinomial and hypergeometric distributions. Poisson, normal, exponential, uniform, geometric, Gamma, Beta, Chi-square, Student, and F-distributions. Method of least squares. Multiple regression.

**Bibliography** .......... 157

**Index** .......... 161

# Chapter 1

# Functions of one variable. Complex numbers

| | | |
|---|---|---|
| 1.1 | $ax^2 + bx + c = 0 \iff x_{1,2} = \dfrac{-b \pm \sqrt{b^2 - 4ac}}{2a}$ | The roots of the general *quadratic* equation. (They are real if $b^2 \geq 4ac$.) |
| 1.2 | $ax^3 + bx^2 + cx + d = 0$ | The general *cubic* equation. |
| 1.3 | $x^3 + px + q = 0$ | (1.2) reduces to the form (1.3) if $x$ in (1.2) is replaced by $x - b/3a$. |
| 1.4 | $x^3 + px + q = 0$ with $\Delta = 4p^3 + 27q^2$ has<br>• three different real roots if $\Delta < 0$;<br>• three real roots, at least two of which are equal, if $\Delta = 0$;<br>• one real and two complex roots if $\Delta > 0$. | Classification of the roots of (1.3). |
| 1.5 | The solutions of (1.3) are<br>$x_1 = u + v, \quad x_2 = -\dfrac{u+v}{2} + \dfrac{u-v}{2}\sqrt{-3},$ and<br>$x_3 = -\dfrac{u+v}{2} - \dfrac{u-v}{2}\sqrt{-3},$ where<br>$u = \sqrt[3]{-\dfrac{q}{2} + \dfrac{1}{2}\sqrt{\dfrac{4p^3 + 27q^2}{27}}}$<br>$v = \sqrt[3]{-\dfrac{q}{2} - \dfrac{1}{2}\sqrt{\dfrac{4p^3 + 27q^2}{27}}}$ | *Cardano's formulas.* |
| 1.6 | $P(x) = a_n x^n + a_{n-1} x^{n-1} + \cdots + a_1 x + a_0$ | A *polynomial* of degree $n$. ($a_n \neq 0$) |

| | | |
|---|---|---|
| 1.7 | For the polynomial in (1.6) there exist constants $x_1, x_2, \ldots, x_n$ (real or complex) such that $$a_n x^n + \cdots + a_1 x + a_0 = a_n(x - x_1)\ldots(x - x_n)$$ | The *fundamental theorem of algebra*. |
| 1.8 | If the coefficients $a_n, a_{n-1}, \ldots, a_1, a_0$ in (1.6) are all integers, $p$ and $q$ are integers without common factors, and $P(p/q) = 0$, then $p$ divides $a_0$ and $q$ divides $a_n$. | *Rational zeros of a polynomial.* |
| 1.9 | Let $k$ be the number of changes of signs in the sequence of coefficients $a_n, a_{n-1}, \ldots, a_1, a_0$ in (1.6). Then the number of positive real zeros of $P(x)$, counting the multiplicities of the roots, is $k$ or $k$ minus a positive even number. If $k = 1$, the equation has just one positive real root. | *Descartes's rule of signs.* |
| 1.10 | The graph of the equation $$Ax^2 + Bxy + Cy^2 + Dx + Ey + F = 0$$ is<br>• an ellipse, a point or empty if $4AC > B^2$;<br>• a parabola, a line, two parallel lines, or empty if $4AC = B^2 \neq 0$;<br>• a hyperbola or two intersecting lines if $4AC < B^2$. | Classification of *conics*. |
| 1.11 | $x = x'\cos\theta - y'\sin\theta, \quad y = x'\sin\theta + y'\cos\theta$ with $\cot 2\theta = (A - C)/B$ | Transforms (1.10) into a quadratic equation in $x'$, $y'$, where the coefficient of $x'y'$ is 0. |
| 1.12 | $(x - x_0)^2 + (y - y_0)^2 = r^2$ | *Circle* with center at $(x_0, y_0)$ and radius $r$. |
| 1.13 | $\dfrac{(x - x_0)^2}{a^2} + \dfrac{(y - y_0)^2}{b^2} = 1, \quad ab \neq 0$ | *Ellipse* with center at $(x_0, y_0)$ and axes parallel to the coordinate axes. |
| 1.14 | $\dfrac{(x - x_0)^2}{a^2} - \dfrac{(y - y_0)^2}{b^2} = \pm 1, \quad ab \neq 0$ | *Hyperbola* with center at $(x_0, y_0)$ and axes parallel to the coordinate axes. |
| 1.15 | Asymptotes of (1.14): $y = \dfrac{a}{b}x$ and $y = -\dfrac{a}{b}x$ | Formulas for asymptotes. |

| | | |
|---|---|---|
| 1.16 | $(x-x_0)^2 = a(y-y_0), \quad a \neq 0$ | Parabola with vertex $(x_0, y_0)$ and axis parallel to the $y$-axis. |
| 1.17 | $d = \sqrt{(x_2-x_1)^2 + (y_2-y_1)^2}$ | The *distance* between the points $(x_1, y_1)$ and $(x_2, y_2)$. |
| 1.18 | The graph of $y = f(x)$ is *symmetric* about the $y$-axis if $f(x) = f(-x)$, about the origin if $f(-x) = -f(x)$, about the line $x = a$ if $f(a+x) = f(a-x)$, and about the point $(a,0)$ if $f(a-x) = -f(a+x)$. | Symmetry properties of the graph of $y = f(x)$. |
| 1.19 | The function $y = f(x)$ is *periodic* with period $k$ if $f(x+k) = f(x)$ for all $x$. The smallest positive period is called the *fundamental period*, or *wave length* of $f$. | Periodic functions. |
| 1.20 | $y = ax + b$ is a *nonvertical asymptote* for the curve $y = f(x)$ if $$\lim_{x \to \infty} (f(x) - (ax+b)) = 0$$ or $$\lim_{x \to -\infty} (f(x) - (ax+b)) = 0$$ | Definition of a nonvertical asymptote. |
| 1.21 | How to find an asymptote for the curve $y = f(x)$ as $x \to \infty$: <br>• Examine $\lim_{x \to \infty} (f(x)/x)$. If the limit does not exist, there is no asymptote as $x \to \infty$. <br>• If $\lim_{x \to \infty} (f(x)/x) = a$, examine $\lim_{x \to \infty} (f(x) - ax)$. If the limit does not exist, there is no asymptote as $x \to \infty$. <br>• If $\lim_{x \to \infty} (f(x) - ax) = b$, $y = ax + b$ is an asymptote for the curve $y = f(x)$ as $x \to \infty$. | Method for finding nonvertical asymptotes for a curve $y = f(x)$ as $x \to \infty$. Replacing $x \to \infty$ by $x \to -\infty$ gives a method for finding nonvertical asymptotes for a curve $y = f(x)$ as $x \to -\infty$. |
| 1.22 | To find an approximate root of $f(x) = 0$, define the sequence $\{x_n\}$ for $n = 1, 2, \ldots,$ by $$x_{n+1} = x_n - \frac{f(x_n)}{f'(x_n)}$$ If $x_0$ is close to an actual root, $x_n$ will usually converge rapidly to that root. | Newton's approximation method. |

1.23 $y - f(x_1) = f'(x_1)(x - x_1)$ | The *tangent* to $y = f(x)$ at $(x_1, f(x_1))$.

1.24 $y - f(x_1) = -\dfrac{1}{f'(x_1)}(x - x_1)$ | The *normal* to $y = f(x)$ at $(x_1, f(x_1))$.

1.25
(i) $a^r \cdot a^s = a^{r+s}$ (ii) $(a^r)^s = a^{rs}$
(iii) $(ab)^r = a^r b^r$ (iv) $a^r/a^s = a^{r-s}$
(v) $\left(\dfrac{a}{b}\right)^r = \dfrac{a^r}{b^r}$

Rules for powers. ($r$ and $s$ are arbitrary real numbers, $a$ and $b$ are positive real numbers.)

1.26 $e = \lim\limits_{n \to \infty} \left(1 + \dfrac{1}{n}\right)^n \approx 2.718281828$ | Definition of $e$. (See also (7.16) for $x = 1$.)

1.27 $e^{\ln x} = x$ | Definition of the natural logarithm.

1.28 $\ln(xy) = \ln x + \ln y$, $\ln \dfrac{x}{y} = \ln x - \ln y$,
$\ln x^p = p \ln x$

Rules for the natural logarithm function. ($x$ and $y$ are positive.)

1.29 $a^{\log_a x} = x$ $(a > 0, a \neq 1)$ | Definition of the *logarithm* with base $a$.

1.30 $\log_a b \cdot \log_b a = 1$, $\log_{10} x = \log_{10} e \cdot \ln x$ | Logarithms with different bases.

1.31 | Definitions of the *trigonometric* functions.

1.32 $1° = \left(\dfrac{\pi}{180}\right)$ rad; $1$ rad $= \left(\dfrac{180}{\pi}\right)°$ | Relationship between degrees and radians (rad).

1.33 $\tan x = \dfrac{\sin x}{\cos x}$, $\cot x = \dfrac{\cos x}{\sin x} = \dfrac{1}{\tan x}$ | Definition of the *tangent* and *cotangent* functions.

1.34

| $x$ | $0$ | $\frac{\pi}{6} = 30°$ | $\frac{\pi}{4} = 45°$ | $\frac{\pi}{3} = 60°$ | $\frac{\pi}{2} = 90°$ |
|---|---|---|---|---|---|
| $\sin x$ | $0$ | $\frac{1}{2}$ | $\frac{1}{2}\sqrt{2}$ | $\frac{1}{2}\sqrt{3}$ | $1$ |
| $\cos x$ | $1$ | $\frac{1}{2}\sqrt{3}$ | $\frac{1}{2}\sqrt{2}$ | $\frac{1}{2}$ | $0$ |
| $\tan x$ | $0$ | $\frac{1}{3}\sqrt{3}$ | $1$ | $\sqrt{3}$ | $*$ |
| $\cot x$ | $*$ | $\sqrt{3}$ | $1$ | $\frac{1}{3}\sqrt{3}$ | $0$ |

\* not defined

Special values of the trigonometric functions.

1.35

| $x$ | $\frac{3\pi}{4} = 135°$ | $\pi = 180°$ | $\frac{3\pi}{2} = 270°$ | $2\pi = 360°$ |
|---|---|---|---|---|
| $\sin x$ | $\frac{1}{2}\sqrt{2}$ | $0$ | $-1$ | $0$ |
| $\cos x$ | $-\frac{1}{2}\sqrt{2}$ | $-1$ | $0$ | $1$ |
| $\tan x$ | $-1$ | $0$ | $*$ | $0$ |
| $\cot x$ | $-1$ | $*$ | $0$ | $*$ |

\* not defined

Trigonometric formulas. (For series expansions of trigonometric functions, see Chapter 7.)

1.36 $\quad \sin^2 x + \cos^2 x = 1$

1.37 $\quad \tan^2 x = \dfrac{1}{\cos^2 x} - 1, \qquad \cot^2 x = \dfrac{1}{\sin^2 x} - 1$

1.38
$\cos(x+y) = \cos x \cos y - \sin x \sin y$
$\cos(x-y) = \cos x \cos y + \sin x \sin y$

1.39
$\sin(x+y) = \sin x \cos y + \cos x \sin y$
$\sin(x-y) = \sin x \cos y - \cos x \sin y$

1.40
$\tan(x+y) = \dfrac{\tan x + \tan y}{1 - \tan x \tan y}$
$\tan(x-y) = \dfrac{\tan x - \tan y}{1 + \tan x \tan y}$

1.41
$\cos 2x = 2\cos^2 x - 1 = 1 - 2\sin^2 x$
$\sin 2x = 2\sin x \cos x$

1.42 $\sin^2 \dfrac{x}{2} = \dfrac{1-\cos x}{2}, \quad \cos^2 \dfrac{x}{2} = \dfrac{1+\cos x}{2}$

Trigonometric formulas. (For series expansions of trigonometric functions, see Chapter 7.)

1.43
$\cos x + \cos y = 2\cos\dfrac{x+y}{2}\cos\dfrac{x-y}{2}$
$\cos x - \cos y = -2\sin\dfrac{x+y}{2}\sin\dfrac{x-y}{2}$

1.44
$\sin x + \sin y = 2\sin\dfrac{x+y}{2}\cos\dfrac{x-y}{2}$
$\sin x - \sin y = 2\cos\dfrac{x+y}{2}\sin\dfrac{x-y}{2}$

1.45 $\sinh x = \dfrac{e^x - e^{-x}}{2}, \quad \cosh x = \dfrac{e^x + e^{-x}}{2}$

*Hyperbolic* sine and cosine.

1.46
$\cosh^2 x - \sinh^2 x = 1$
$\cosh(x+y) = \cosh x \cosh y + \sinh x \sinh y$
$\cosh 2x = \cosh^2 x + \sinh^2 x$
$\sinh(x+y) = \sinh x \cosh y + \cosh x \sinh y$
$\sinh 2x = 2 \sinh x \cosh x$

Properties of hyperbolic functions.

## Complex numbers

1.47 $z = a + ib, \quad \bar{z} = a - ib$

A *complex number* and its *conjugate*. $a$ and $b$ are real numbers, and $i^2 = -1$.

1.48 $|z| = \sqrt{a^2 + b^2}$

The *modulus* (or *magnitude*) of $z = a + ib$.

1.49 $|\bar{z}_1| = |z_1|, \; z_1\bar{z}_1 = |z_1|^2, \; \overline{z_1 + z_2} = \bar{z}_1 + \bar{z}_2,$
$|z_1 z_2| = |z_1||z_2|, \; |z_1 + z_2| \le |z_1| + |z_2|$

Basic rules. $z_1$ and $z_2$ are complex numbers.

1.50 $(a+bi)(c+di) = (ac-bd) + (ad+bc)i$

*Multiplication* of complex numbers.

1.51 $\dfrac{c+di}{a+bi} = \dfrac{1}{a^2+b^2}\big((ac+bd) + (ad-bc)i\big)$

*Division* of complex numbers.

| | | |
|---|---|---|
| 1.52 | $z = a + ib = r(\cos\theta + i\sin\theta)$, where $$\cos\theta = \frac{a}{\sqrt{a^2+b^2}}, \quad \sin\theta = \frac{b}{\sqrt{a^2+b^2}}$$ | The *polar* or *trigonometric* form of a complex number. |
| 1.53 | If $z_k = r_k(\cos\theta_k + i\sin\theta_k)$, $k = 1, 2$, then $$z_1 z_2 = r_1 r_2 \big(\cos(\theta_1+\theta_2) + i\sin(\theta_1+\theta_2)\big)$$ $$\frac{z_1}{z_2} = \frac{r_1}{r_2}\big(\cos(\theta_1-\theta_2) + i\sin(\theta_1-\theta_2)\big)$$ | Multiplication and division on trigonometric form. |
| 1.54 | $(\cos\theta + i\sin\theta)^n = \cos n\theta + i\sin n\theta$ | *De Moivre's formula*, $n = 0, 1, \ldots$ |
| 1.55 | If $z = x + iy$, then $e^z = e^{x+iy} = e^x \cdot e^{iy} = e^x(\cos y + i\sin y)$ | The *complex exponential function*. |
| 1.56 | $e^{\bar{z}} = \overline{e^z}$, $\quad e^{z+2\pi i} = e^z$, $\quad e^{z_1+z_2} = e^{z_1} e^{z_2}$, $\quad e^{z_1-z_2} = e^{z_1}/e^{z_2}$ | Rules for the complex exponential function. |
| 1.57 | $\cos z = \dfrac{e^{iz}+e^{-iz}}{2}, \quad \sin z = \dfrac{e^{iz}-e^{-iz}}{2i}$ | *Euler's formulas*. |
| 1.58 | If $a = r(\cos\theta + i\sin\theta)$, then the equation $$z^n = a$$ has exactly $n$ roots, namely $$z_k = \sqrt[n]{r}\Big(\cos\frac{\theta+2k\pi}{n} + i\sin\frac{\theta+2k\pi}{n}\Big)$$ for $k = 0, 1, \ldots, n-1$. | *nth roots* of a complex number, $n = 1, 2, \ldots$ |

## References

Most of the formulas are found in any calculus text, e.g. Edwards and Penney (1990). For (1.3)–(1.5) and (1.9), see e.g. Dickson (1939).

# Chapter 2

# Limits. Differentiation (one variable)

| | | |
|---|---|---|
| 2.1 | $f(x)$ tends to $A$ as a *limit* as $x$ approaches $a$, $\lim_{x \to a} f(x) = A$ or $f(x) \to A$ as $x \to a$, if for every number $\varepsilon > 0$ there exists a number $\delta > 0$ such that $\|f(x) - A\| < \varepsilon$ whenever $0 < \|x - a\| < \delta$ | The definition of a limit of a function of one variable. |
| 2.2 | If $\lim_{x \to a} f(x) = A$ and $\lim_{x \to a} g(x) = B$, then<br>• $\lim_{x \to a} (f(x) + g(x)) = A + B$<br>• $\lim_{x \to a} (f(x) - g(x)) = A - B$<br>• $\lim_{x \to a} (f(x) \cdot g(x)) = A \cdot B$<br>• $\lim_{x \to a} \dfrac{f(x)}{g(x)} = \dfrac{A}{B}$ (if $B \neq 0$) | Rules for limits. |
| 2.3 | $f$ is *continuous* at $x = a$ if $\lim_{x \to a} f(x) = f(a)$ | Definition of continuity of a function of one variable. |
| 2.4 | If $f$ and $g$ are continuous at $a$, then<br>• $f \pm g$ and $f \cdot g$ are continuous at $a$.<br>• $f/g$ is continuous at $a$ if $g(a) \neq 0$. | Properties of continuous functions. |
| 2.5 | If $g$ is continuous at $a$, and $f$ is continuous at $g(a)$, then $f(g(x))$ is continuous at $a$. | Continuity of *compositions*. |
| 2.6 | Any function built from continuous functions by additions, subtractions, multiplications, divisions, and compositions, is continuous where defined. | A useful result. |
| 2.7 | If $f$ is continuous on an interval $I$ containing $a$ and $b$, and $A$ lies between $f(a)$ and $f(b)$, then there is a $\xi$ between $a$ and $b$ such that $A = f(\xi)$. | The *intermediate value theorem*. |

2.8 $\quad f'(x) = \lim\limits_{h \to 0} \dfrac{f(x+h) - f(x)}{h}$ | The definition of the derivative. If the limit exists, $f$ is called *differentiable* at $x$.

If $y = f(x)$, other notations for the derivative include

2.9 $\quad f'(x) = y' = \dfrac{dy}{dx} = \dfrac{df(x)}{dx} = Df(x)$ | Other notations for the derivative.

2.10 $\quad y = f(x) \pm g(x) \;\Rightarrow\; y' = f'(x) \pm g'(x)$ | General rules.

2.11 $\quad y = f(x)g(x) \;\Rightarrow\; y' = f'(x)g(x) + f(x)g'(x)$

2.12 $\quad y = \dfrac{f(x)}{g(x)} \;\Rightarrow\; y' = \dfrac{f'(x)g(x) - f(x)g'(x)}{\bigl(g(x)\bigr)^2}$

2.13 $\quad y = f\bigl(g(x)\bigr) \;\Rightarrow\; y' = f'(g(x)) \cdot g'(x)$

2.14 $\quad y = \bigl(f(x)\bigr)^{g(x)} \;\Rightarrow\;$
$\qquad y' = \bigl(f(x)\bigr)^{g(x)} \left(g'(x)\ln f(x) + g(x)\dfrac{f'(x)}{f(x)}\right)$

2.15 $\quad y = f^{-1}(x) \;\Rightarrow\; y' = \dfrac{1}{f'\bigl(f^{-1}(x)\bigr)}$ | $f^{-1}$ is the *inverse* of a one-to-one function $f$.

2.16 $\quad y = c \quad (c \text{ constant}) \;\Rightarrow\; y' = 0$ | Special rules.

2.17 $\quad y = x^a \quad (a \text{ constant}) \;\Rightarrow\; y' = ax^{a-1}$

2.18 $\quad y = \dfrac{1}{x} \;\Rightarrow\; y' = -\dfrac{1}{x^2}$

2.19 $\quad y = \sqrt{x} \;\Rightarrow\; y' = \dfrac{1}{2\sqrt{x}}$

2.20 $\quad y = e^x \;\Rightarrow\; y' = e^x$

2.21 $\quad y = a^x \;\Rightarrow\; y' = a^x \ln a, \quad a > 0$

2.22 $\quad y = \ln x \;\Rightarrow\; y' = \dfrac{1}{x}$

2.23 $\quad y = \log_a x \;\Rightarrow\; y' = \dfrac{1}{x}\log_a e, \quad a > 0,\; a \neq 1$

| | | |
|---|---|---|
| 2.24 | $y = \sin x \Rightarrow y' = \cos x$ | Special rules. |
| 2.25 | $y = \cos x \Rightarrow y' = -\sin x$ | |
| 2.26 | $y = \tan x \Rightarrow y' = \dfrac{1}{\cos^2 x} = 1 + \tan^2 x$ | |
| 2.27 | $y = \cot x \Rightarrow y' = -\dfrac{1}{\sin^2 x} = -(1 + \cot^2 x)$ | |
| 2.28 | $y = \sin^{-1} x = \arcsin x \Rightarrow y' = \dfrac{1}{\sqrt{1-x^2}}$ | |
| 2.29 | $y = \cos^{-1} x = \arccos x \Rightarrow y' = -\dfrac{1}{\sqrt{1-x^2}}$ | |
| 2.30 | $y = \tan^{-1} x = \arctan x \Rightarrow y' = \dfrac{1}{1+x^2}$ | |
| 2.31 | $y = \cot^{-1} x = \text{arccot}\, x \Rightarrow y' = -\dfrac{1}{1+x^2}$ | |
| 2.32 | $y = \sinh x \Rightarrow y' = \cosh x$ | |
| 2.33 | $y = \cosh x \Rightarrow y' = \sinh x$ | |
| 2.34 | If $f$ is continuous in $[a,b]$ and differentiable in $(a,b)$, then there exists at least one point $\xi$ in $(a,b)$ such that $$f'(\xi) = \frac{f(b) - f(a)}{b - a}$$ | The *mean value theorem*. |
| 2.35 | If $f$ and $g$ are continuous in $[a,b]$ and differentiable in $(a,b)$, then there exists at least one point $\xi$ in $(a,b)$ such that $$[f(b) - f(a)]g'(\xi) = [g(b) - g(a)]f'(\xi)$$ | The *generalized mean value theorem*. |
| 2.36 | Suppose $f$ and $g$ are differentiable in an interval $(\alpha, \beta)$ around $a$, except possibly at $a$, and suppose that $f(x)$ and $g(x)$ both approach 0 when $x$ approaches $a$. If $g'(x) \neq 0$ for all $x \neq a$ in $(\alpha, \beta)$ and $\lim_{x \to a} f'(x)/g'(x) = L$ ($L$ finite, $L = \infty$ or $L = -\infty$), then $$\lim_{x \to a} \frac{f(x)}{g(x)} = \lim_{x \to a} \frac{f'(x)}{g'(x)} = L$$ | *L'Hôpital's rule*. The same rule applies if $x \to a^+$, $x \to a^-$, $x \to \infty$, $x \to -\infty$, and if $f(x) \to \pm\infty$, $g(x) \to \pm\infty$. |

2.37  If $y = f(x)$ and $dx$ is any number,
$$dy = f'(x)\, dx$$
is the *differential* of $y$.

Definition of the differential.

2.38  $f(x+dx) - f(x) \approx f'(x)\, dx$  when $dx$ is small

A useful approximation, made more precise in (2.39).

2.39  $f(x+dx) - f(x) = f'(x)\, dx + \varepsilon\, dx$
where $\varepsilon \to 0$ as $dx \to 0$

Property of a differentiable function. (If $dx$ is very small, then $\varepsilon$ is very small, and $\varepsilon\, dx$ is "very, very small".)

2.40
$$d(af + bg) = a\, df + b\, dg \quad (a \text{ and } b \text{ constants})$$
$$d(fg) = g\, df + f\, dg$$
$$d(f/g) = (g\, df - f\, dg)/g^2$$
$$df(u) = f'(u)\, du$$

Rules for differentials. $f$ and $g$ are differentiable, and $u$ is any differentiable function.

## References

All formulas are standard and are found in almost any calculus text, e.g. Edwards and Penney (1990).

# Chapter 3

# Partial derivatives

3.1 If $z = f(x_1, \ldots, x_n) = f(\mathbf{x})$, then
$$\frac{\partial f}{\partial x_i} = D_{x_i} f = \frac{\partial z}{\partial x_i} = f_i = f'_i(\mathbf{x})$$
all denote the derivative of $f(x_1, \ldots, x_n)$ with respect to $x_i$ when all the other variables are held constant.

Definition of the *partial derivative*. (Other notations may be used.)

3.2 $$\frac{\partial^2 z}{\partial x_j \partial x_i} = f''_{ij}(x_1, \ldots, x_n) = \frac{\partial}{\partial x_j} f'_i(x_1, \ldots, x_n)$$

Second order partial derivatives of $z = f(x_1, \ldots, x_n)$.

3.3 $$\frac{\partial^2 f}{\partial x_j \partial x_i} = \frac{\partial^2 f}{\partial x_i \partial x_j}, \quad i, j = 1, 2, \ldots, n$$

*Young's (or Schwarz's) theorem*, valid if one of the two partials is continuous.

3.4 $f(x_1, \ldots, x_n)$ is said to be *of class $C^k$*, or only $C^k$, in the set $S \subset R^n$ if all partial derivatives of $f$ of order $\leq k$ are continuous in $S$.

Definition of a $C^k$-function.

3.5 $z = F(x, y)$, $x = f(t)$, $y = g(t)$ $\Rightarrow$
$$\frac{dz}{dt} = F'_1(x, y)\frac{dx}{dt} + F'_2(x, y)\frac{dy}{dt}$$

The *chain rule*.

3.6 If $z = F(x_1, \ldots, x_n)$, $x_i = f_i(t_1, \ldots, t_m)$, $i = 1, \ldots, n$, then for all $j = 1, \ldots, m$,
$$\frac{\partial z}{\partial t_j} = \sum_{i=1}^{n} \frac{\partial F(x_1, \ldots, x_n)}{\partial x_i} \frac{\partial x_i}{\partial t_j}$$

The chain rule. (General case.)

3.7 If $z = f(x_1, \ldots, x_n)$ and $dx_1, \ldots, dx_n$ are arbitrary numbers,
$$dz = \sum_{i=1}^{n} f'_i(x_1, \ldots, x_n) dx_i$$
is the *differential* of $z$.

Definition of the *differential*.

3.8 $\Delta z \approx dz$ when $dx_1, \ldots, dx_n$ are all small, where
$\Delta z = f(x_1 + dx_1, \ldots, x_n + dx_n) - f(x_1, \ldots, x_n)$

| A useful approximation, made more precise for differentiable functions in (3.9).

3.9 $f$ is *differentiable* at **x** if $f'_i(\mathbf{x})$ all exist and there exist functions $\varepsilon_i = \varepsilon_i(dx_1, \ldots, dx_n)$, $i = 1, \ldots, n$, that all approach zero as $dx_i$ all approach zero and such that
$$\Delta z - dz = \varepsilon_1 dx_1 + \cdots + \varepsilon_n dx_n$$

| Definition of differentiability.

3.10 If $f$ is a $C^1$-function, i.e. has continuous first order partials, then $f$ is differentiable.

| An important fact.

3.11
$$d(af + bg) = a\,df + b\,dg \quad (a \text{ and } b \text{ constants})$$
$$d(fg) = g\,df + f\,dg$$
$$d(f/g) = (g\,df - f\,dg)/g^2$$
$$dF(u) = F'(u)\,du$$

| Rules for differentials. $f$ and $g$ are differentiable functions of $x_1, \ldots, x_n$, $F$ is a differentiable function of one variable, and $u$ is any differentiable function of $x_1, \ldots, x_n$.

3.12 $F(x, y) = c \Rightarrow \dfrac{dy}{dx} = -\dfrac{F'_1(x, y)}{F'_2(x, y)}, \quad F'_2(x, y) \neq 0$

| Formula for the slope of a level curve for $z = F(x, y)$. For precise assumptions, see (3.14).

3.13 If $y = f(x)$ is a $C^2$-function satisfying $F(x, y) = c$, then
$$f''(x) = -\frac{[F''_{11}(F'_2)^2 - 2F''_{12}F'_1 F'_2 + F''_{22}(F'_1)^2]}{(F'_2)^3}$$
$$= \frac{1}{(F'_2)^3} \begin{vmatrix} 0 & F'_1 & F'_2 \\ F'_1 & F''_{11} & F''_{12} \\ F'_2 & F''_{12} & F''_{22} \end{vmatrix}$$

| A useful result. All partials are evaluated at $(x, y)$.

3.14 If $F(x, y)$ is $C^k$ in a set $A$, $(x_0, y_0)$ is an interior point of $A$, $F(x_0, y_0) = c$, and $F'_2(x_0, y_0) \neq 0$, then $F(x, y) = c$ defines $y$ as a $C^k$-function of $x$, $y = \varphi(x)$, in some neighborhood of $(x_0, y_0)$, and
$$\frac{dy}{dx} = -\frac{F'_1(x, y)}{F'_2(x, y)}$$

| The *implicit function theorem*. (For a more general result, see (5.3).)

3.15 If $F(x_1, x_2, \ldots, x_n, z) = c$, (c constant), then
$$\frac{\partial z}{\partial x_i} = -\frac{\partial F/\partial x_i}{\partial F/\partial z}, \quad \frac{\partial F}{\partial z} \neq 0, \quad i = 1, 2, \ldots, n$$

A generalization of (3.12).

3.16 $f(\mathbf{x}) = f(x_1, \ldots, x_n)$ is *homogeneous of degree* $k$ in $D$ if for all $t > 0$ and all $\mathbf{x} \in D$,
$$f(tx_1, tx_2, \ldots, tx_n) = t^k f(x_1, x_2, \ldots, x_n)$$

The definition of a homogeneous function. We assume that $D$ is a *cone* in the sense that $\mathbf{x} \in D$ and $t > 0$ implies $t\mathbf{x} \in D$.

3.17 $f(\mathbf{x}) = f(x_1, \ldots, x_n)$ is homogeneous of degree $k$ in the open cone $D$ if and only if
$$\sum_{i=1}^{n} x_i f'_i(\mathbf{x}) = k f(\mathbf{x})$$

The *Euler theorem*, valid for $C^1$-functions.

3.18 If $f(\mathbf{x}) = f(x_1, \ldots, x_n)$ is homogeneous of degree $k$ in the open cone $D$, then
- $\partial f / \partial x_i$ is homogeneous of degree $k - 1$
- $\sum_{i=1}^{n} \sum_{j=1}^{n} x_i x_j f''_{ij}(\mathbf{x}) = k(k-1) f(\mathbf{x})$

Properties of homogeneous functions. (A cone is defined in (3.16).)

3.19 $f(\mathbf{x}) = f(x_1, \ldots, x_n)$ is *homothetic* in the cone $D$ if for all $\mathbf{x}, \mathbf{y} \in D$ and all $t > 0$,
$$f(\mathbf{x}) = f(\mathbf{y}) \Rightarrow f(t\mathbf{x}) = f(t\mathbf{y})$$

Definition of homothetic function. (A cone is defined in (3.16).)

3.20 Let $f(\mathbf{x})$ be a continuous, homothetic function defined in the convex cone $D$. Assume that $f$ is strictly increasing along each ray in $D$ (i.e. for each $\mathbf{x}_0 \in D$, $f(t\mathbf{x}_0)$ is a strictly increasing function of $t$). Then there exists a homogeneous function $g$ and a strictly increasing function $F$ such that
$$f(\mathbf{x}) = F(g(\mathbf{x})) \text{ for all } \mathbf{x} \in D$$

A property of continuous, homothetic functions (which is sometimes taken as the definition of homotheticity). One can assume that $g$ is homogeneous of degree 1.

3.21 $\nabla f(\mathbf{x}) = \left( \dfrac{\partial f(\mathbf{x})}{\partial x_1}, \ldots, \dfrac{\partial f(\mathbf{x})}{\partial x_n} \right)$

The *gradient* of $f$ at $\mathbf{x} = (x_1, \ldots, x_n)$.

3.22 $f'_\mathbf{a}(\mathbf{x}) = \lim\limits_{h \to 0} \dfrac{f(\mathbf{x} + h\mathbf{a}) - f(\mathbf{x})}{h}, \quad \|\mathbf{a}\| = 1$

The *directional derivative* of $f$ at $\mathbf{x}$ in the direction of $\mathbf{a}$.

3.23 $f'_\mathbf{a}(\mathbf{x}) = \sum\limits_{i=1}^{n} f'_i(\mathbf{x}) a_i = \nabla f(\mathbf{x}) \cdot \mathbf{a}$

The relationship between the directional derivative and the gradient.

| | | |
|---|---|---|
| 3.24 | - $\nabla f(\mathbf{x})$ is orthogonal to the level surface $f(\mathbf{x}) = C$.<br>- $\nabla f(\mathbf{x})$ points in the direction of maximal increase of $f$.<br>- $\|\nabla f(\mathbf{x})\|$ measures the rate of change of $f$ in the direction of $\nabla f(\mathbf{x})$. | Properties of the gradient. |
| 3.25 | The *tangent plane* to $z = f(x,y)$ at the point $(x_0, y_0, z_0)$, with $z_0 = f(x_0, y_0)$, has the equation $$z - z_0 = f_1'(x_0, y_0)(x-x_0) + f_2'(x_0, y_0)(y-y_0)$$ | Definition of a tangent plane. |
| 3.26 | The *tangent hyperplane* to $F(\mathbf{x}) = F(x_1, \ldots, x_n) = C$ at the point $\mathbf{x}^0 = (x_1^0, \ldots, x_n^0)$ has the equation $$\nabla F(\mathbf{x}^0) \cdot (\mathbf{x} - \mathbf{x}^0) = 0$$ | Definition of the tangent hyperplane. The vector $\nabla F(\mathbf{x}^0)$ is a *normal* to the hyperplane. |

## References

Most of the formulas are found in any calculus text, e.g. Edwards and Penney (1990). For properties of homothetic functions, see Shephard (1970) and Førsund (1975).

# Chapter 4

# Elasticities. Elasticities of substitution

4.1 $\quad \text{El}_x f(x) = \dfrac{x}{f(x)} f'(x) = \dfrac{d(\ln f(x))}{d(\ln x)}$ 
$\quad$ $\text{El}_x f(x)$, the *elasticity* of $f(x)$ with respect to $x$, is (approximately) the percentage change in $f(x)$ corresponding to a one per cent increase in $x$.

4.2 $\quad \text{El}_x(f(x)g(x)) = \text{El}_x f(x) + \text{El}_x g(x)$
$\quad$ General rules for calculating elasticities.

4.3 $\quad \text{El}_x \left( \dfrac{f(x)}{g(x)} \right) = \text{El}_x f(x) - \text{El}_x g(x)$

4.4 $\quad \text{El}_x(f(x) + g(x)) = \dfrac{f(x)\,\text{El}_x f(x) + g(x)\,\text{El}_x g(x)}{f(x) + g(x)}$

4.5 $\quad \text{El}_x(f(x) - g(x)) = \dfrac{f(x)\,\text{El}_x f(x) - g(x)\,\text{El}_x g(x)}{f(x) - g(x)}$

4.6 $\quad \text{El}_x f(g(x)) = \text{El}_u f(u)\,\text{El}_x u, \quad u = g(x)$

4.7 $\quad \text{El}_x A = 0, \quad \text{El}_x x^a = a, \quad \text{El}_x e^x = x$
$\quad$ Special rules for elasticities.

4.8 $\quad \text{El}_x \sin x = x \cot x, \quad \text{El}_x \cos x = -x \tan x$

4.9 $\quad \text{El}_x \tan x = \dfrac{x}{\sin x \cos x}, \quad \text{El}_x \cot x = \dfrac{-x}{\sin x \cos x}$

4.10 $\quad \text{El}_x \ln x = 1/\ln x, \quad \text{El}_x \log_a x = 1/\ln x$

4.11 $\quad \text{El}_i f(\mathbf{x}) = \dfrac{x_i}{f(\mathbf{x})} \dfrac{\partial f(\mathbf{x})}{\partial x_i}, \quad i = 1, \ldots, n$
$\quad$ The *partial elasticity* of $f(\mathbf{x}) = f(x_1, \ldots, x_n)$ wrt. $x_i$.

4.12 If $z = F(x_1, \ldots, x_n)$ and $x_i = f_i(t_1, \ldots, t_m)$ for $i = 1, \ldots, n$, then for $j = 1, \ldots, m$,
$$\mathrm{El}_{t_j} z = \sum_{i=1}^{n} \mathrm{El}_i F(x_1, \ldots, x_n) \mathrm{El}_{t_j} x_i$$

The *chain rule for elasticities*.

4.13 The *directional elasticity* of $f$ at $\mathbf{x}$, in the direction of $\mathbf{x}/\|\mathbf{x}\|$, is
$$\mathrm{El}_{\mathbf{a}} f(\mathbf{x}) = \frac{\|\mathbf{x}\|}{f(\mathbf{x})} f'_{\mathbf{a}}(\mathbf{x}), \quad \mathbf{a} = \frac{\mathbf{x}}{\|\mathbf{x}\|}$$

$\mathrm{El}_{\mathbf{a}} f(\mathbf{x})$ is approximately the percentage change in $f(\mathbf{x})$ corresponding to a one per cent increase in each component of $\mathbf{x}$. (See also (4.14). $f'_{\mathbf{a}}(\mathbf{x})$ is defined in (3.22).)

4.14 $$\mathrm{El}_{\mathbf{a}} f(\mathbf{x}) = \sum_{i=1}^{n} \mathrm{El}_i f(\mathbf{x}), \quad \mathbf{a} = \frac{\mathbf{x}}{\|\mathbf{x}\|}$$

A useful fact (the *passus equation*).

4.15 $$R_{yx} = \frac{f'_1(x,y)}{f'_2(x,y)}, \quad f(x,y) = c$$

The *marginal rate of substitution* between $y$ and $x$, $R_{yx}$, is (approximately) how much we must add of $y$ per unit of $x$ removed to stay on the same level curve.

4.16
- When $f$ is a utility function and $x$ and $y$ are goods, or when $f$ is a production function and $x$ and $y$ are inputs, $R_{yx}$ is called the *marginal rate of substitution* (abbreviated MRS).
- When $f(x,y) = 0$ is a production function on implicit form (for given factor inputs), and $x$ and $y$ are two different products, $R_{yx}$ is called the *marginal rate of product transformation* (abbreviated MRPT).

Different special cases of (4.15).

4.17 $$\sigma_{yx} = \mathrm{El}_{R_{yx}}\left(\frac{y}{x}\right) = -\frac{\partial \ln\left(\frac{y}{x}\right)}{\partial \ln\left(\frac{f'_2}{f'_1}\right)}, \quad f(x,y) = c$$

The *elasticity of substitution* between $y$ and $x$, $\sigma_{yx}$, is (approximately) the percentage change in the factor ratio $y/x$ corresponding to a one percent change in the marginal rate of substitution.

| | | |
|---|---|---|
| 4.18 | $\sigma_{yx} = \mathrm{El}_{R_{yx}}\left(\dfrac{y}{x}\right) = -\dfrac{\partial \ln\left(\dfrac{y}{x}\right)}{\partial \ln\left(\dfrac{p_2}{p_1}\right)}, \quad f(x,y) = c$ | The MRTS assuming competitive behavior, with $p_1$ and $p_2$ as the factor prices. |
| 4.19 | $\sigma_{yx} = \dfrac{\dfrac{1}{xf_1'} + \dfrac{1}{yf_2'}}{-\dfrac{f_{11}''}{(f_1')^2} + 2\dfrac{f_{12}''}{f_1'f_2'} - \dfrac{f_{22}''}{(f_2')^2}}, \quad f(x,y) = c$ | An alternative formula for the elasticity of substitution. |
| 4.20 | If $f(x,y)$ is homogeneous of degree 1, then $\sigma_{yx} = \dfrac{f_1'f_2'}{ff_{12}''}$ | A special case. |
| 4.21 | $\sigma_{ij} = -\dfrac{\partial \ln\left(\dfrac{x_i}{x_j}\right)}{\partial \ln\left(\dfrac{f_i'}{f_j'}\right)}, \quad f(x_1,\ldots,x_n) = c,\ i \neq j$ | The elasticity of substitution in the $n$ variable case. |
| 4.22 | $\sigma_{ij} = \dfrac{\dfrac{1}{x_if_i'} + \dfrac{1}{x_jf_j'}}{-\dfrac{f_{ii}''}{(f_i')^2} + \dfrac{2f_{ij}''}{f_i'f_j'} - \dfrac{f_{jj}''}{(f_j')^2}}, \quad i \neq j$ | The elasticity of substitution, $f(x_1,\ldots,x_n) = c$. |
| 4.23 | $\sigma_{ij} = -\dfrac{\partial \ln\left(\dfrac{C_i'(\mathbf{w},y)}{C_j'(\mathbf{w},y)}\right)}{\partial \ln\left(\dfrac{w_i}{w_j}\right)}, \quad i \neq j$ $y$, $C$ and $w_k$ (for $k \neq i,j$) are constants. | The shadow elasticity of substitution in production theory between factor $i$ and $j$. $C$ is the cost function, and $\mathbf{w}$ is the vector of factor prices. See Chapter 25. |
| 4.24 | $\sigma_{ij} = \dfrac{-\dfrac{C_{ii}''}{(C_i')^2} + \dfrac{2C_{ij}''}{C_i'C_j'} - \dfrac{C_{jj}''}{(C_j')^2}}{\dfrac{1}{w_iC_i'} + \dfrac{1}{w_jC_j'}}, \quad i \neq j$ | An alternative form of (4.23). |

4.25 $A_{ij}(\mathbf{w}, y) = \dfrac{C(\mathbf{w},y)C''_{ij}(\mathbf{w},y)}{C'_i(\mathbf{w},y)C'_j(\mathbf{w},y)}, \quad i \neq j$

The *Allen-Uzawa elasticity of substitution* in production theory. $C(\mathbf{w}, y)$ is the cost function, $\mathbf{w} = (w_1, \ldots, w_n)$ is the vector of input prices, $y$ is output. (See (25.2).) (For a discussion of this concept, see Blackorby and Russell (1989).)

4.26 $A_{ij}(\mathbf{w}, y) = \dfrac{\varepsilon_{ij}(\mathbf{w},y)}{S_j(\mathbf{w},y)}, \quad i \neq j$

Here $\varepsilon_{ij}(\mathbf{w}, y)$ is the (constant-output) cross-price elasticity of demand and $S_j(\mathbf{w}, y) = p_j C_j(\mathbf{w}, y)/C(\mathbf{w}, y)$ is the share of the $j$th input in total cost.

4.27 $\begin{aligned} M_{ij}(\mathbf{w}, y) &= \dfrac{w_i C''_{ij}(\mathbf{w}, y)}{C'_j(\mathbf{w}, y)} - \dfrac{w_i C''_{ii}(\mathbf{w}, y)}{C'_i(\mathbf{w}, y)} \\ &= \varepsilon_{ji}(\mathbf{w}, y) - \varepsilon_{ii}(\mathbf{w}, y), \quad i \neq j \end{aligned}$

The *Morishima elasticity of substitution* in production theory.

4.28 If $n > 2$, then $M_{ij}(\mathbf{w}, y) = M_{ji}(\mathbf{w}, y)$ for all $i \neq j$ if and only if all $M_{ij}(\mathbf{w}, y)$ are equal to one and the same constant.

Symmetry of the Morishima elasticity of substitution.

## References

These formulas are usually not found in calculus texts. (4.2)–(4.10) follow by applying definition (4.1). (4.12) follows from (3.5) and (4.11). For (4.21)–(4.28) see Blackorby and Russell (1989) and Fuss and McFadden (1978).

# Chapter 5

# Systems of equations

5.1
$$f_1(x_1, x_2, \ldots, x_n, y_1, y_2, \ldots, y_m) = 0$$
$$f_2(x_1, x_2, \ldots, x_n, y_1, y_2, \ldots, y_m) = 0$$
$$\cdots\cdots\cdots\cdots\cdots\cdots\cdots\cdots\cdots\cdots\cdots$$
$$f_m(x_1, x_2, \ldots, x_n, y_1, y_2, \ldots, y_m) = 0$$

A general system of equations with $n$ exogenous variables, $x_1, \ldots, x_n$, and $m$ endogenous variables, $y_1, \ldots, y_m$.

5.2
$$\frac{\partial \mathbf{f}(\mathbf{x}, \mathbf{y})}{\partial \mathbf{y}} = \begin{pmatrix} \frac{\partial f_1}{\partial y_1} & \cdots & \frac{\partial f_1}{\partial y_m} \\ \vdots & \ddots & \vdots \\ \frac{\partial f_m}{\partial y_1} & \cdots & \frac{\partial f_m}{\partial y_m} \end{pmatrix}$$

The *Jacobian matrix* of $f_1, \ldots, f_m$ with respect to $y_1, \ldots, y_m$.

5.3
Suppose $f_1, \ldots, f_m$ are $C^k$-functions in a set $A$ in $R^{n+m}$, let $(\mathbf{x}^0, \mathbf{y}^0) = (x_1^0, \ldots, x_n^0, y_1^0, \ldots, y_m^0)$ be a solution to (5.1) in the interior of $A$. Suppose also that the determinant of the Jacobian matrix $\partial \mathbf{f}(\mathbf{x}, \mathbf{y})/\partial \mathbf{y}$ in (5.2) is different from 0 at $(\mathbf{x}^0, \mathbf{y}^0)$. Then (5.1) defines $y_1, \ldots, y_m$ as $C^k$-functions of $x_1, \ldots, x_n$ in some neighborhood of $(\mathbf{x}^0, \mathbf{y}^0)$, and in that neighborhood, for $j = 1, \ldots, n$,

$$\begin{pmatrix} \frac{\partial y_1}{\partial x_j} \\ \vdots \\ \frac{\partial y_m}{\partial x_j} \end{pmatrix} = - \begin{pmatrix} \frac{\partial f_1}{\partial y_1} & \cdots & \frac{\partial f_1}{\partial y_m} \\ \vdots & \ddots & \vdots \\ \frac{\partial f_m}{\partial y_1} & \cdots & \frac{\partial f_m}{\partial y_m} \end{pmatrix}^{-1} \begin{pmatrix} \frac{\partial f_1}{\partial x_j} \\ \vdots \\ \frac{\partial f_m}{\partial x_j} \end{pmatrix}$$

The *general implicit function theorem.* (It gives sufficient conditions for system (5.1) to define the endogenous variables $y_1, \ldots, y_m$ as differentiable functions of the exogenous variables $x_1, \ldots, x_n$.

5.4
$$f_1(x_1, x_2, \ldots, x_n) = 0$$
$$f_2(x_1, x_2, \ldots, x_n) = 0$$
$$\ldots\ldots\ldots\ldots\ldots\ldots\ldots$$
$$f_m(x_1, x_2, \ldots, x_n) = 0$$

A general system of $m$ equations and $n$ variables.

5.5 System (5.4) has $k$ *degrees of freedom* if there is a set of $k$ of the variables that can be freely chosen such that the remaining $n-k$ variables are uniquely determined when the $k$ variables have been assigned specific values. If the variables are restricted to vary in a set $S$ in $R^n$, the system has $k$ degrees of freedom in $S$.

Definition of degrees of freedom for a system of equations.

5.6 To find the number of degrees of freedom for a system of equations, count the number, $n$, of variables and the number, $m$, of equations. If $n > m$, there are $n - m$ degrees of freedom in the system. If $n < m$, there is, in general, no solution of the system.

The "counting rule". This is a rough rule which is *not* valid in general.

5.7 If the conditions in (5.3) are satisfied, then system (5.1) has $n$ degrees of freedom.

A precise (local) counting rule.

5.8 $$\mathbf{f}'(\mathbf{x}) = \begin{pmatrix} \dfrac{\partial f_1(\mathbf{x})}{\partial x_1} & \cdots & \dfrac{\partial f_1(\mathbf{x})}{\partial x_n} \\ \vdots & & \vdots \\ \dfrac{\partial f_m(\mathbf{x})}{\partial x_1} & \cdots & \dfrac{\partial f_m(\mathbf{x})}{\partial x_n} \end{pmatrix}$$

The *Jacobian matrix* of $f_1, \ldots, f_m$ with respect to $x_1, \ldots, x_n$, also denoted by $\partial \mathbf{f}(\mathbf{x})/\partial \mathbf{x}$. (See (5.2).)

5.9 If $\mathbf{x}^0 = (x_1^0, \ldots, x_n^0)$ is a solution of (5.4), $m \leq n$, and the rank of the Jacobian matrix $\mathbf{f}'(\mathbf{x})$ is equal to $m$, then system (5.4) has $n-m$ degrees of freedom in some neighborhood of $\mathbf{x}^0$.

A precise (local) counting rule. (Valid if $f_1, \ldots, f_m$ are $C^1$.)

5.10 The functions $f_1(\mathbf{x}), \ldots, f_m(\mathbf{x})$ are *functionally dependent* in a set $A$ in $R^n$ if there exists a real-valued $C^1$-function $F$ defined on a neighborhood of $S = \{(f_1(\mathbf{x}), \ldots, f_m(\mathbf{x})) : \mathbf{x} \in A\}$ such that
$$F(f_1(\mathbf{x}), \ldots, f_m(\mathbf{x})) = 0 \quad \text{for all } \mathbf{x} \in A$$
where $\nabla F \neq \mathbf{0}$ in $S$.

Definition of functional dependence.

| | | |
|---|---|---|
| 5.11 | If $f_1(\mathbf{x}), \ldots, f_m(\mathbf{x})$ are functionally dependent and $m \leq n$, then the rank of $J(\mathbf{x})$ is less than $m$. | A necessary condition for functional dependence. |
| 5.12 | If (5.4) has solutions, and if $f_1(\mathbf{x}), \ldots, f_m(\mathbf{x})$ are functionally dependent, then system (5.4) has at least one redundant equation. | A sufficient condition for the counting rule to fail. |
| 5.13 | $\det(\mathbf{f}'(\mathbf{x})) = \begin{vmatrix} \dfrac{\partial f_1(\mathbf{x})}{\partial x_1} & \cdots & \dfrac{\partial f_1(\mathbf{x})}{\partial x_n} \\ \vdots & \ddots & \vdots \\ \dfrac{\partial f_n(\mathbf{x})}{\partial x_1} & \cdots & \dfrac{\partial f_n(\mathbf{x})}{\partial x_n} \end{vmatrix}$ | The *Jacobian determinant* of $f_1, \ldots, f_n$ with respect to $x_1, \ldots, x_n$. |
| 5.14 | If $f_1(\mathbf{x}), \ldots, f_n(\mathbf{x})$ are functionally dependent, then $\det(\mathbf{f}(\mathbf{x})) \equiv 0$. | A special case of (5.11). The converse is not generally true. |
| 5.15 | $\begin{aligned} y_1 &= f_1(x_1, \ldots, x_n) \\ &\cdots\cdots\cdots\cdots\cdots \\ y_n &= f_n(x_1, \ldots, x_n) \end{aligned} \iff \mathbf{y} = \mathbf{f}(\mathbf{x})$ | A transformation $\mathbf{f}$ from $R^n$ to $R^n$. |
| 5.16 | Suppose the transformation $\mathbf{f}$ in (5.15) is $C^1$ in a neighborhood of $\mathbf{x}^0$ and that the Jacobian determinant in (5.13) is not zero at $\mathbf{x}^0$. Then there exists a $C^1$ transformation $\mathbf{g}$ which is locally an inverse to $\mathbf{f}$, i.e. $\mathbf{g}(\mathbf{f}(\mathbf{x})) = \mathbf{x}$ for all $\mathbf{x}$ in some neighborhood of $\mathbf{x}^0$. | The *local inverse function theorem*. |
| 5.17 | Suppose $\mathbf{f} : R^n \to R^n$ is $C^1$ and that there exist numbers $h$ and $k$ such that for all $\mathbf{x}$ and all $i, j = 1, \ldots, n$, $\|\det(\mathbf{f}'(\mathbf{x}))\| \geq h > 0$ and $|\partial f_i(\mathbf{x})/\partial x_j| \leq k$ Then $\mathbf{f}$ has an inverse defined and $C^1$ on all of $R^n$. | A *global inverse function theorem*. (Hadamard.) |
| 5.18 | Suppose $\mathbf{f} : R^n \to R^n$ is $C^1$ and that the determinant in (5.13) is $\neq 0$ for all $\mathbf{x}$. Then $\mathbf{f}(\mathbf{x})$ has an inverse defined and $C^1$ on all of $R^n$ if and only if $\inf\{\|\mathbf{f}(\mathbf{x})\| : \|\mathbf{x}\| \geq n\} \to \infty$ when $n \to \infty$ | A *global inverse function theorem*. |

| | | |
|---|---|---|
| 5.19 | Suppose $\mathbf{f}: R^n \to R^n$ is $C^1$ and let $\Omega$ be the rectangle $\Omega = \{\mathbf{x} \in R^n : \mathbf{a} \leq \mathbf{x} \leq \mathbf{b}\}$, where $\mathbf{a}$ and $\mathbf{b}$ are given vectors in $R^n$. Then $\mathbf{f}$ is one to one in $\Omega$ if *one* of the following conditions are satisfied for all $\mathbf{x}$:<br>• The Jacobian matrix $\mathbf{f}'(\mathbf{x})$ has only strictly positive principal minors.<br>• The Jacobian matrix $\mathbf{f}'(\mathbf{x})$ has only strictly negative principal minors. | A *Gale-Nikaido theorem*. |
| 5.20 | An $n \times n$-matrix $\mathbf{A}$ (not necessarily symmetric) is called *positive quasi-definite* if for each $n$-vector $\mathbf{x} \neq \mathbf{0}$, $\mathbf{x}'\mathbf{A}\mathbf{x} > 0$. | Definition of a positive quasi-definite matrix. |
| 5.21 | Suppose $\mathbf{f}: R^n \to R^n$ is a $C^1$ function and assume that the Jacobian matrix $\mathbf{f}'(\mathbf{x})$ is positive quasi-definite everywhere in a convex set $\Omega$. Then $\mathbf{f}$ is one-to-one in $\Omega$. | A *Gale-Nikaido theorem*. |
| 5.22 | $\mathbf{f}: R^n \to R^n$ is a *contraction mapping* if there exists a constant $k \in [0, 1)$ such that for all $\mathbf{x} \in R^n$, $\mathbf{y} \in R^n$,<br>$$\|\mathbf{f}(\mathbf{x}) - \mathbf{f}(\mathbf{y})\| \leq k\|\mathbf{x} - \mathbf{y}\|$$ | Definition of a contraction mapping. |
| 5.23 | If $\mathbf{f}: R^n \to R^n$ is a *contraction mapping*, then $\mathbf{f}$ has a unique *fixed point*, i.e. a point $\mathbf{x}^* \in R^n$ such that $\mathbf{f}(\mathbf{x}^*) = \mathbf{x}^*$.<br>For any $\mathbf{x}^0 \in R^n$ we have $\mathbf{x}^* = \lim_{n \to \infty} \mathbf{f}(\mathbf{x}_n)$, where $\mathbf{x}_n = \mathbf{f}(\mathbf{x}_{n-1})$ for $n = 1, 2, \ldots$ | The existence of a fixed point for a contraction map. (This result can be generalized to "complete metric spaces". See Bartle (1976).) |
| 5.24 | Let $K$ be a compact, convex set in $R^n$ and $\mathbf{f}$ a continuous function mapping $K$ into $K$. Then $\mathbf{f}$ has a fixed point $\mathbf{x}^* \in K$, i.e. a point $\mathbf{x}^*$ such that $\mathbf{f}(\mathbf{x}^*) = \mathbf{x}^*$. | *Brouwer's fixed point theorem*. |
| 5.25 | Let $K$ be a compact, convex set in $R^n$ and $\mathbf{f}$ a correspondence which to each point $\mathbf{x}$ in $K$ associates a nonempty, convex subset $\mathbf{f}(\mathbf{x})$ of $K$. Suppose that $\mathbf{f}$ has a closed graph, i.e. the set<br>$$\{(\mathbf{x}, \mathbf{y}) \in R^{2n} : \mathbf{x} \in K \text{ and } \mathbf{y} \in \mathbf{f}(\mathbf{x})\}$$<br>is closed in $R^{2n}$. Then $\mathbf{f}$ has a fixed point, i.e. a point $\mathbf{x}^* \in K$, such that $\mathbf{x}^* \in \mathbf{f}(\mathbf{x}^*)$. | *Kakutani's fixed point theorem*. |

5.26
$$\begin{aligned} a_{11}x_1 + a_{12}x_2 + \cdots + a_{1n}x_n &= b_1 \\ a_{21}x_1 + a_{22}x_2 + \cdots + a_{2n}x_n &= b_2 \\ &\cdots \\ a_{m1}x_1 + a_{m2}x_2 + \cdots + a_{mn}x_n &= b_m \end{aligned}$$

The *general linear system of equations with $m$ equations and $n$ unknowns*.

5.27
$$\mathbf{A} = \begin{pmatrix} a_{11} & \cdots & a_{1n} \\ a_{21} & \cdots & a_{2n} \\ \vdots & & \vdots \\ a_{m1} & \cdots & a_{mn} \end{pmatrix}$$

$$\mathbf{A_b} = \begin{pmatrix} a_{11} & \cdots & a_{1n} & b_1 \\ a_{21} & \cdots & a_{2n} & b_2 \\ \vdots & & \vdots & \vdots \\ a_{m1} & \cdots & a_{mn} & b_m \end{pmatrix}$$

$\mathbf{A}$ is the coefficient matrix of (5.26), while $\mathbf{A_b}$ is the "extended coefficient matrix" of (5.26).

5.28  System (5.26) has a solution $\iff$ r($\mathbf{A}$)=r($\mathbf{A_b}$).

5.29  If r($\mathbf{A}$) = r($\mathbf{A_b}$) = $k < m$, then system (5.26) has $m - k$ superfluous equations.

Main results about linear systems of equations. r($\mathbf{B}$) denotes the rank of the matrix $\mathbf{B}$. (See (19.22).)

5.30  If r($\mathbf{A}$) = r($\mathbf{A_b}$) = $k < n$, then system (5.26) has $n - k$ degrees of freedom.

## References

For (5.1)–(5.16) see e.g. Rudin (1982). For (5.17)–(5.21) see Parthasarathy (1983). For (5.22)–(5.23) see Bartle (1976). For (5.24)–(5.25) see Nikaido (1970) or Scarf (1973). (5.26)–(5.30) are standard results in linear algebra, see e.g. Anton (1987).

# Chapter 6

# Inequalities

| | | |
|---|---|---|
| 6.1 | $\lvert \lvert a \rvert - \lvert b \rvert \rvert \leq \lvert a \pm b \rvert \leq \lvert a \rvert + \lvert b \rvert$ | Triangle inequalities. $a$ and $b$ are real (or complex) numbers. |
| 6.2 | $\dfrac{n}{\sum_{i=1}^{n} 1/a_i} \leq \left( \prod_{i=1}^{n} a_i \right)^{1/n} \leq \dfrac{\sum_{i=1}^{n} a_i}{n}$ | Harmonic mean $\leq$ Geometric mean $\leq$ Arithmetic mean. Equalities if and only if $a_1 = \cdots = a_n$. |
| 6.3 | $\dfrac{2}{1/a_1 + 1/a_2} \leq \sqrt{a_1 a_2} \leq \dfrac{a_1 + a_2}{2}$ | (6.2) for $n = 2$. |
| 6.4 | $a_1^{\lambda_1} \ldots a_n^{\lambda_n} \leq \lambda_1 a_1 + \cdots + \lambda_n a_n$ | Inequality for *weighted means*. $a_i \geq 0$, $\sum_{i=1}^{n} \lambda_i = 1, \lambda_i \geq 0$. |
| 6.5 | $a_1^{\lambda} a_2^{1-\lambda} \leq \lambda a_1 + (1-\lambda) a_2$ | (6.4) for $n = 2$, $a_1 \geq 0$, $a_2 \geq 0$, $\lambda \in [0,1]$. |
| 6.6 | $\sum_{i=1}^{n} \lvert a_i b_i \rvert \leq \left( \sum_{i=1}^{n} \lvert a_i \rvert^p \right)^{1/p} \left( \sum_{i=1}^{n} \lvert b_i \rvert^q \right)^{1/q}$ | *Hölder's inequality* for sums. $1/p + 1/q = 1$, $p > 1$, $q > 1$. Equality if and only if $\lvert b_i \rvert = c \lvert a_i \rvert^{p-1}$ for some positive constant $c$. |
| 6.7 | $\left( \sum_{i=1}^{n} \lvert a_i b_i \rvert \right)^2 \leq \left( \sum_{i=1}^{n} a_i^2 \right) \left( \sum_{i=1}^{n} b_i^2 \right)$ | *Cauchy-Schwarz's inequality* for sums. (Put $p = q = 2$ in (6.6).) |
| 6.8 | $\left( \sum_{i=1}^{n} a_i \right) \left( \sum_{i=1}^{n} b_i \right) \leq n \sum_{i=1}^{n} a_i b_i$ | *Chebychev's inequality*. $a_1 \geq \cdots \geq a_n$ $b_1 \geq \cdots \geq b_n$ |

6.9 $$\left(\sum_{i=1}^{n}|a_i+b_i|^p\right)^{1/p} \leq \left(\sum_{i=1}^{n}|a_i|^p\right)^{1/p} + \left(\sum_{i=1}^{n}|b_i|^p\right)^{1/p}$$

Minkowski's inequality for sums. $p \geq 1$, $a_i > 0$, $b_i > 0$. Equality if and only if $|b_i| = c|a_i|$ for some positive constant $c$.

6.10 If $f$ is convex,
$$f\left(\sum_{i=1}^{n} a_i x_i\right) \leq \sum_{i=1}^{n} a_i f(x_i)$$

Jensen's inequality for sums. $\sum_{i=1}^{n} a_i = 1$, $a_i \geq 0$, $i = 1, \ldots, n$.

6.11 $$\left(\sum_{i=1}^{n}|a_i|^q\right)^{1/q} \leq \left(\sum_{i=1}^{n}|a_i|^p\right)^{1/p}$$

Another Jensen's inequality; $0 < p < q$.

6.12 $$\int_a^b |f(x)g(x)|\,dx \leq \left(\int_a^b |f(x)|^p\,dx\right)^{1/p} \left(\int_a^b |g(x)|^q\,dx\right)^{1/q}$$

Hölder's inequality for integrals. Equality holds if and only if $|g(x)| = c|f(x)|^{p-1}$ for some positive constant $c$. $1/p + 1/q = 1$, $p > 1$, $q > 1$.

6.13 $$\left(\int_a^b f(x)g(x)\,dx\right)^2 \leq \int_a^b (f(x))^2\,dx \int_a^b (g(x))^2\,dx$$

Cauchy-Schwarz's inequality for integrals.

6.14 $$\left(\int_a^b |f(x)+g(x)|^p\,dx\right)^{1/p} \leq \left(\int_a^b |f(x)|^p\,dx\right)^{1/p} + \left(\int_a^b |g(x)|^p\,dx\right)^{1/p}$$

Minkowski's inequality for integrals. $p > 1$. Equality holds if and only if $g(x) = cf(x)$ for some positive constant $c$.

6.15 If $f$ is convex, then
$$f\left(\int a(x)g(x)\,dx\right) \leq \int a(x)f(g(x))\,dx$$

Jensen's inequality for integrals. $\int a(x)\,dx = 1$, $a(x) \geq 0$, $f(u) \geq 0$. $f$ must be defined on the range of $g$.

6.16 If $f$ is convex on the interval $I$ and $X$ is a random variable with finite expectation, then
$$f(E(X)) \leq E(f(X))$$
If $f$ is strictly convex, the inequality is strict unless $X$ is a constant with probability 1.

Special case of Jensen's inequality. $E$ is the expectation operator.

6.17 If $U$ is concave on the interval $I$ and $X$ is a random variable with finite expectation, then
$$E(U(X)) \leq U(E(X))$$
An important fact in utility theory. (It follows from (6.16) by putting $f = -U$.)

## References

The best reference to inequalities is Hardy, Littlewood, and Pólya (1952).

# Chapter 7

# Series. Taylor formulas

| | | |
|---|---|---|
| 7.1 | $\sum_{i=0}^{n-1}(a+id) = na + \dfrac{n(n-1)d}{2}$ | Sum of the first $n$ terms of an *arithmetic series*. |
| 7.2 | $a + ak + ak^2 + ak^3 + \cdots = a\dfrac{1-k^n}{1-k}, \quad k \neq 1$ | Sum of the first $n$ terms of a *geometric series*. |
| 7.3 | $a + ak + ak^2 + ak^3 + \cdots = \dfrac{a}{1-k} \quad \text{if } |k| < 1$ | Sum of an infinite geometric series. |
| 7.4 | $a_1 + a_2 + \cdots + a_n + \cdots = s$ means that $\lim_{n\to\infty}(a_1 + a_2 + \cdots + a_n) = s$ | Definition of the *convergence* of an infinite series. |
| 7.5 | $a_1 + \cdots + a_n + \cdots$ converges $\Rightarrow \lim_{n\to\infty} a_n = 0$ | A necessary (but NOT sufficient) condition for the convergence of an infinite series. |
| 7.6 | $\lim_{n\to\infty} \left|\dfrac{a_{n+1}}{a_n}\right| < 1 \Rightarrow \sum_{n=1}^{\infty} a_n$ converges | The *ratio test*. |
| 7.7 | $\lim_{n\to\infty} \left|\dfrac{a_{n+1}}{a_n}\right| > 1 \Rightarrow \sum_{n=1}^{\infty} a_n$ diverges | The *ratio test*. |
| 7.8 | If $\sum a_n$ is a series with only positive terms and $f(x)$ is a positive-valued, decreasing, continuous function for $x \geq 1$, and if $f(n) = a_n$ for all integers $n \geq 1$, then the infinite series and the improper integral $$\sum_{n=1}^{\infty} a_n \quad \text{and} \quad \int_1^{\infty} f(x)\,dx$$ either both converge or both diverge. | The *integral test*. |

If $0 \leq a_n \leq b_n$ for all $n$, then

7.9
- $\sum a_n$ converges if $\sum b_n$ converges.
- $\sum b_n$ diverges if $\sum a_n$ diverges.

*The comparison test.*

7.10 $\displaystyle\sum_{n=1}^{\infty} \frac{1}{n^p}$ is convergent $\iff p > 1$

*An important result.*

7.11
$$f(x) = f(0) + f'(0)\frac{x}{1!} + \cdots + f^{(n)}(0)\frac{x^n}{n!}$$
$$+ f^{(n+1)}(\theta x)\frac{x^{n+1}}{(n+1)!}, \quad 0 < \theta < 1$$

*Maclaurin's formula, with error term.*

7.12
$$f(a+x) = f(a) + f'(a)\frac{x}{1!} + \cdots + f^{(n)}(a)\frac{x^n}{n!}$$
$$+ f^{(n+1)}(a+\theta x)\frac{x^{n+1}}{(n+1)!}, \quad 0 < \theta < 1$$

*Taylor's formula, with error term.*

7.13 $f(x) = f(0) + f'(0)\dfrac{x}{1!} + f''(0)\dfrac{x^2}{2!} + \cdots$

Maclaurin's series for $f(x)$, valid for those $x$ where the error term in (7.11) goes to 0 as $n$ goes to $\infty$.

7.14 $f(a+x) = f(a) + f'(a)\dfrac{x}{1!} + f''(a)\dfrac{x^2}{2!} + \cdots$

Taylor series for $f(x+a)$, valid for those $x$ where the error term in (7.12) goes to 0 as $n$ goes to $\infty$.

7.15
$$f(\mathbf{x}^0 + \mathbf{h}) =$$
$$f(\mathbf{x}^0) + \sum_{i=1}^{n} f'_i(\mathbf{x}^0)h_i + \tfrac{1}{2}\sum_{i=1}^{n}\sum_{j=1}^{n} f''_{ij}(\mathbf{x}^0 + \theta\mathbf{h})h_i h_j$$

Taylor's formula for functions of several variables, $\theta \in (0,1)$.

7.16 $e^x = 1 + \dfrac{x}{1!} + \dfrac{x^2}{2!} + \dfrac{x^3}{3!} + \cdots$

Valid for all $x$.

7.17 $\ln(1+x) = x - \dfrac{x^2}{2} + \dfrac{x^3}{3} - \dfrac{x^4}{4} + \cdots$

Valid for $-1 < x \leq 1$.

7.18 $(1+x)^m = \dbinom{m}{0} + \dbinom{m}{1}x + \dbinom{m}{2}x^2 + \cdots$

Valid for $-1 < x < 1$. For the definition of $\binom{m}{k}$, see (7.23).

7.19 $\quad \sin x = x - \dfrac{x^3}{3!} + \dfrac{x^5}{5!} - \dfrac{x^7}{7!} + \dfrac{x^9}{9!} - \cdots$ $\qquad$ Valid for all $x$.

7.20 $\quad \cos x = 1 - \dfrac{x^2}{2!} + \dfrac{x^4}{4!} - \dfrac{x^6}{6!} + \dfrac{x^8}{8!} - \cdots$ $\qquad$ Valid for all $x$.

7.21 $\quad \arcsin x = x + \dfrac{1}{2}\dfrac{x^3}{3} + \dfrac{1\cdot 3}{2\cdot 4}\dfrac{x^5}{5} + \dfrac{1\cdot 3\cdot 5}{2\cdot 4\cdot 6}\dfrac{x^7}{7} + \cdots$ $\qquad$ Valid for $|x| \leq 1$.

7.22 $\quad \arctan x = x - \dfrac{x^3}{3} + \dfrac{x^5}{5} - \dfrac{x^7}{7} + \cdots$ $\qquad$ Valid for $|x| \leq 1$.

7.23 $\quad \dbinom{m}{k} = \dfrac{m(m-1)\cdots(m-k+1)}{k!}, \quad \dbinom{m}{0} = 1$ $\qquad$ *Binomial coefficients.* ($m$ is an arbitrary real number, $k$ is a natural number.)

7.24 $\quad \displaystyle\sum_{i=0}^{n} \binom{n}{i} a^{n-i} b^{i} = (a+b)^n$ $\qquad$ *Newton's binomial formula.*

7.25 $\quad \displaystyle\sum_{i=0}^{n} \binom{n}{i} = (1+1)^n = 2^n$ $\qquad$ A special case of (7.24).

7.26 $\quad \displaystyle\sum_{i=r}^{n} \binom{i}{r} = \binom{n+1}{r+1}$ $\qquad$ Properties of binomial coefficients.

7.27 $\quad \displaystyle\sum_{i=0}^{j} \binom{n}{i}\binom{m}{j-i} = \binom{n+m}{j}$

7.28 $\quad \displaystyle\sum_{i=0}^{n} \binom{n}{i}^2 = \binom{2n}{n}$

7.29 $\quad 1 + 3 + 5 + \cdots + (2n-1) = n^2$ $\qquad$ Summation formulas.

7.30 $\quad 1 + 2 + 3 + \cdots + n = \dfrac{n(n+1)}{2}$

7.31 $\quad 1^2 + 2^2 + 3^2 + \cdots + n^2 = \dfrac{n(n+1)(2n+1)}{6}$

7.32 $\quad 1^3 + 2^3 + 3^3 + \cdots + n^3 = \left(\dfrac{n(n+1)}{2}\right)^2$

7.33 $$1^4 + 2^4 + 3^4 + \cdots + n^4 = \frac{n(n+1)(2n+1)(3n^2+3n-1)}{30}$$ | Summation formula.

## References

All formulas are standard and are found in most calculus texts, e.g. Edwards and Penney (1990).

# Chapter 8

# Integration

**Indefinite integrals**

| | | |
|---|---|---|
| 8.1 | $\int f(x)\,dx = F(x) + C \Leftrightarrow F'(x) = f(x)$ | Definition of the indefinite integral. $f$ is continuous. |
| 8.2 | $\int (af(x) + bg(x))\,dx = a\int f(x)\,dx + b\int g(x)\,dx$ | Linearity of the integral. $a$ and $b$ are constants. |
| 8.3 | $\int f(x)g'(x)\,dx = f(x)g(x) - \int f'(x)g(x)\,dx$ | Integration by parts. |
| 8.4 | $\int f(x)\,dx = \int f(g(t))g'(t)\,dt, \quad x = g(t)$ | Change of variable. (Integration by substitution.) |
| 8.5 | $\int x^n\,dx = \begin{cases} \dfrac{x^{n+1}}{n+1} + C, & n \neq -1 \\ \ln|x| + C, & n = -1 \end{cases}$ | Special integration results. |
| 8.6 | $\int a^x\,dx = \dfrac{1}{\ln a}a^x + C, \quad a > 0$ | |
| 8.7 | $\int e^x\,dx = e^x + C$ | |
| 8.8 | $\int xe^x\,dx = xe^x - e^x + C$ | |
| 8.9 | $\int x^n e^{ax}\,dx = \dfrac{x^n}{a}e^{ax} - \dfrac{n}{a}\int x^{n-1}e^{ax}\,dx, \quad a \neq 0$ | |
| 8.10 | $\int \log_a x\,dx = x\log_a x - x\log_a e + C, \quad a > 0$ | |

8.11 $\int \ln x \, dx = x \ln x - x + C$ — Special integration results.

8.12 $\int x^n \ln x \, dx = \dfrac{x^{n+1}\big((n+1)\ln x - 1\big)}{(n+1)^2} + C$ — $(n \neq -1)$

8.13 $\int \sin x \, dx = -\cos x + C$

8.14 $\int \cos x \, dx = \sin x + C$

8.15 $\int \tan x \, dx = -\ln|\cos x| + C$

8.16 $\int \cot x \, dx = \ln|\sin x| + C$

8.17 $\int \dfrac{1}{\sin x} \, dx = \ln\left|\dfrac{1-\cos x}{\sin x}\right| + C$

8.18 $\int \dfrac{1}{\cos x} \, dx = \ln\left|\dfrac{1+\sin x}{\cos x}\right| + C$

8.19 $\int \dfrac{1}{\sin^2 x} \, dx = -\cot x + C$

8.20 $\int \dfrac{1}{\cos^2 x} \, dx = \tan x + C$

8.21 $\int \sin^2 x \, dx = \tfrac{1}{2}x - \tfrac{1}{2}\sin x \cos x + C$

8.22 $\int \cos^2 x \, dx = \tfrac{1}{2}x + \tfrac{1}{2}\sin x \cos x + C$

8.23 $\int \sin^n x \, dx = -\dfrac{\sin^{n-1} x \cos x}{n} + \dfrac{n-1}{n}\int \sin^{n-2} x \, dx$ — $(n \neq 0)$

8.24 $\int \cos^n x \, dx = \dfrac{\cos^{n-1} x \sin x}{n} + \dfrac{n-1}{n}\int \cos^{n-2} x \, dx$ — $(n \neq 0)$

8.25 $\int e^{\alpha x} \sin \beta x \, dx = \dfrac{e^{\alpha x}}{\alpha^2 + \beta^2}(\alpha \sin \beta x - \beta \cos \beta x) + C$ — $(\alpha^2 + \beta^2 \neq 0)$

8.26 $\int e^{\alpha x} \cos \beta x \, dx =$

$\dfrac{e^{\alpha x}}{\alpha^2 + \beta^2} (\beta \sin \beta x + \alpha \cos \beta x) + C$

Special integration results. $(\alpha^2 + \beta^2 \neq 0)$

8.27 $\int \dfrac{1}{x^2 - a^2} \, dx = \dfrac{1}{2a} \ln \left| \dfrac{x-a}{x+a} \right| + C$ $\quad (a \neq 0)$

8.28 $\int \dfrac{1}{x^2 + a^2} \, dx = \dfrac{1}{a} \arctan \dfrac{x}{a} + C$ $\quad (a \neq 0)$

8.29 $\int \dfrac{1}{\sqrt{a^2 - x^2}} \, dx = \arcsin \dfrac{x}{a} + C$ $\quad (a \neq 0)$

8.30 $\int \dfrac{1}{\sqrt{x^2 \pm a^2}} \, dx = \ln \left| x + \sqrt{x^2 \pm a^2} \right| + C$

8.31 $\int \sqrt{a^2 - x^2} \, dx = \dfrac{x}{2} \sqrt{a^2 - x^2} + \dfrac{a^2}{2} \arcsin \dfrac{x}{a} + C$ $\quad (a \neq 0)$

8.32 $\int \sqrt{x^2 \pm a^2} \, dx =$

$\dfrac{x}{2} \sqrt{x^2 \pm a^2} \pm \dfrac{a^2}{2} \ln \left| x + \sqrt{x^2 \pm a^2} \right| + C$

8.33 $\int \dfrac{dx}{ax^2 + 2bx + c} =$

$\dfrac{1}{2\sqrt{b^2 - ac}} \ln \left| \dfrac{ax + b - \sqrt{b^2 - ac}}{ax + b + \sqrt{b^2 - ac}} \right| + C$

$(b^2 > ac, a \neq 0)$

8.34 $\int \dfrac{dx}{ax^2 + 2bx + c} =$

$\dfrac{1}{\sqrt{ac - b^2}} \arctan \dfrac{ax + b}{\sqrt{ac - b^2}} + C$

$(b^2 < ac, a \neq 0)$

8.35 $\int \dfrac{dx}{ax^2 + 2bx + c} = \dfrac{-1}{ax + b} + C$ $\quad (b^2 = ac, a \neq 0)$

## Definite integrals

| | | |
|---|---|---|
| 8.36 | $\int_a^b f(x)\,dx = F(x)\big|_a^b = F(b) - F(a)$ for $F$ such that $F'(x) = f(x)$ for all $x \in [a,b]$ | Definition of the *definite integral* of a continuous function. |
| 8.37 | $\int_a^b f(x)\,dx = -\int_b^a f(x)\,dx$ <br> $\int_a^a f(x)\,dx = 0$ <br> $\int_a^b \alpha f(x)\,dx = \alpha \int_a^b f(x)\,dx$ <br> $\int_a^b f(x)\,dx = \int_a^c f(x)\,dx + \int_c^b f(x)\,dx$ | $a$, $b$, $c$, and $\alpha$ are arbitrary real numbers. |
| 8.38 | $\int_a^b f(g(x))g'(x)\,dx = \int_{g(a)}^{g(b)} f(u)\,du, \quad u = g(x)$ | *Change of variable.* |
| 8.39 | $\int_a^b f(x)g'(x)\,dx = f(x)g(x)\big|_a^b - \int_a^b f'(x)g(x)\,dx$ | *Integration by parts.* |
| 8.40 | $\int_a^\infty f(x)\,dx = \lim_{M\to\infty} \int_a^M f(x)\,dx$ | Provided the limit exists. |
| 8.41 | $\int_{-\infty}^b f(x)\,dx = \lim_{N\to\infty} \int_{-N}^b f(x)\,dx$ | Provided the limit exists. |
| 8.42 | $\int_{-\infty}^\infty f(x)\,dx =$ <br> $\lim_{N\to\infty} \int_{-N}^a f(x)\,dx + \lim_{M\to\infty} \int_a^M f(x)\,dx$ | *Both* limits on the right-hand side must exist, and $a$ is an arbitrary number. |
| 8.43 | $\int_a^b f(x)\,dx = \lim_{h\to 0+} \int_{a+h}^b f(x)\,dx$ | The definition of the integral if $f$ is continuous in $(a,b]$. |
| 8.44 | $\int_a^b f(x)\,dx = \lim_{h\to 0+} \int_a^{b-h} f(x)\,dx$ | The definition of the integral if $f$ is continuous in $[a,b)$. |
| 8.45 | $|f(x)| \leq g(x)$ for all $x \geq a \Rightarrow$ <br> $\left|\int_a^\infty f(x)\,dx\right| \leq \int_a^\infty g(x)\,dx$ | *Comparison test* for integrals. $f$ and $g$ are continuous for $x \geq a$. |
| 8.46 | $\dfrac{d}{dx}\int_a^x f(t)\,dt = f(x)$ | An important fact. |
| 8.47 | $\dfrac{d}{dx}\int_x^b f(t)\,dt = -f(x)$ | Another important fact. |

8.48 $\dfrac{d}{dx}\int_a^b f(x,t)\,dt = \int_a^b f'_x(x,t)\,dt$ — $a$ and $b$ are independent of $x$.

8.49 $\dfrac{d}{dx}\int_c^\infty f(x,t)\,dt = \int_c^\infty f'_x(x,t)\,dt$ — For precise assumptions see e.g. Bartle (1976), section 33.

8.50 $(d/dx)\int_{u(x)}^{v(x)} f(x,t)\,dt =$
$f(x,v(x))v'(x) - f(x,u(x))u'(x) + \int_{u(x)}^{v(x)} f'_x(x,t)\,dt$ — *Leibniz's formula.*

8.51 $\Gamma(x) = \int_0^\infty e^{-t} t^{x-1}\,dt, \quad x > 0$ — The *Gamma function.*

8.52 $\Gamma(x+1) = x\,\Gamma(x) \quad$ for all $\quad x > 0$ — The *functional equation* for the Gamma function.

8.53 $\int_{-\infty}^{+\infty} e^{-at^2}\,dt = \sqrt{\pi/a} \quad (a>0)$ — An important formula.

8.54 $\int_0^\infty t^k e^{-at^2}\,dt = \dfrac{1}{2} a^{-(k+1)/2} \Gamma((k+1)/2)$ — Valid for $a>0,\ k>-1$.

8.55 $\Gamma(x) = \sqrt{2\pi}\, x^{x-\frac{1}{2}} e^{-x} e^{\theta/12x}, \quad x>0,\ \theta \in (0,1)$ — *Stirling's formula.*

8.56 $B(p,q) = \int_0^1 u^{p-1}(1-u)^{q-1}\,du, \quad p>0, q>0$ — The *Beta function.*

8.57 $B(p,q) = \dfrac{\Gamma(p)\Gamma(q)}{\Gamma(p+q)}$ — The relationship between the Beta function and the Gamma function.

8.58 $\int_a^b f(x)\,dx \approx$
$\dfrac{b-a}{2n}[y_0 + 2y_1 + 2y_2 + \cdots + 2y_{n-1} + y_n]$ — The *trapezoid formula.* $y_i = f(a + i\dfrac{b-a}{n})$.

8.59 $\int_a^b f(x)\,dx \approx$
$\dfrac{b-a}{6n}\left[y_0 + 4\sum_{i=1}^{n} y_{2i-1} + 2\sum_{i=1}^{n-1} y_{2i} + y_{2n}\right]$ — *Simpson's formula.* $2n$ is the even number of equal subintervals in the partition of $[a,b]$. $y_i = f(a + i\dfrac{b-a}{2n})$.

## Multiple integrals

8.60 $$\iint_R f(x,y)\,dx\,dy = \int_a^b (\int_c^d f(x,y)\,dy)\,dx$$
$$= \int_c^d (\int_a^b f(x,y)\,dx)\,dy$$

Definition of the *double integral* of $f(x,y)$ over a rectangle $R = [a,b] \times [c,d]$.

8.61 $$V = \int_a^b (\int_{u(x)}^{v(x)} f(x,y)\,dy)\,dx$$

The double integral of $f(x,y)$ over the region in figure A.

8.62 $$V = \int_c^d (\int_{p(y)}^{q(y)} f(x,y)\,dx)\,dy$$

The double integral of $f(x,y)$ over the region in figure B.

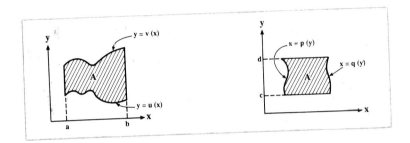

A            B

8.63 $F''_{xy}(x,y) = f(x,y), \quad (x,y) \in [a,b] \times [c,d] \Rightarrow$
$$\int_c^d (\int_a^b f(x,y)\,dx)\,dy =$$
$F(b,d) - F(a,d) - F(b,c) + F(a,c)$

Fundamental theorem.

8.64 $$\iint_A f(x,y)\,dx\,dy =$$
$$\iint_{A'} f(g(u,v), h(u,v)) \left| \frac{\partial(g,h)}{\partial(u,v)} \right| du\,dv$$

New variables in the double integral. $x = g(u,v)$, $y = h(u,v)$ is a one-to-one $C^1$-transformation of $A'$ onto $A$, and the Jacobian $\partial(g,h)/\partial(u,v)$ does not vanish. $f$ is continuous.

8.65 
$$\iint \cdots \int_\Omega f(\mathbf{x})\,dx_1 \cdots dx_{n-1}\,dx_n =$$
$$\int_{a_n}^{b_n} \int_{a_{n-1}}^{b_{n-1}} \cdots \int_{a_1}^{b_1} f(\mathbf{x})\,dx_1) \cdots dx_{n-1})\,dx_n$$

The $n$-*integral* of $f$ over an $n$-dimensional rectangle $\Omega$. $\mathbf{x} = (x_1, \ldots, x_n)$.

8.66
$$\int \cdots \int_A f(\mathbf{x})\,dx_1 \cdots dx_n =$$
$$\int \cdots \int_{A'} f(g_1(\mathbf{u}), \ldots, g_n(\mathbf{u}))\,|J|\,du_1 \cdots du_n$$

New variables in the $n$-integral. $x_i = g_i(\mathbf{u})$, $i = 1, \ldots, n$, is a one-to-one $C^1$-transformation of $A'$ onto $A$, and $J = \dfrac{\partial(g_1, \ldots, g_n)}{\partial(u_1, \ldots, u_n)}$ (the Jacobian) does not vanish. $f$ is continuous.

## References

Most of the formulas are found in any calculus text, e.g. Edwards and Penney (1990). For (8.51)–(8.57), see e.g. Bartle (1976).

# Chapter 9

# Difference equations. (Recurrence relations)

| | | |
|---|---|---|
| 9.1 | $x_{t+1} = a_t x_t + b_t, \quad t = 0, 1, \ldots$ | A *first order linear difference equation*. |
| 9.2 | $x_t = \left(\prod_{k=0}^{t-1} a_k\right) x_0 + \sum_{s=0}^{t-1} \left(\prod_{k=s+1}^{t-1} a_k\right) b_s$ | The solution of (9.1) for $t = 1, 2, \ldots$, if we define $\prod_{k=t}^{t-1} a_k$ to be 1. |
| 9.3 | $x_t = a^t x_0 + \sum_{s=0}^{t-1} a^{t-s-1} b_s, \quad t = 1, 2, \ldots$ | The solution of (9.1) when $a_t = a$, a constant. |
| 9.4 | $x_{t+1} = ax_t + b \Leftrightarrow x_t = a^t\left(x_0 - \dfrac{b}{1-a}\right) + \dfrac{b}{1-a}$ | Equation (9.1) and its solution when $a_t = a \neq 1$, $b_t = b$. |
| 9.5 | $x_{t+n} + a_1(t)x_{t+n-1} + \cdots + a_n(t)x_t = b(t)$ | The *general linear difference equation of order $n$*. |
| 9.6 | $x_{t+n} + a_1(t)x_{t+n-1} + \cdots + a_n(t)x_t = 0$ | The *homogeneous* equation associated with (9.5). |
| 9.7 | The general solution of (9.6) is $$x_t = C_1 u_1(t) + \cdots + C_n u_n(t)$$ where $u_1(t), \ldots, u_n(t)$ are linearly independent solutions of (9.6) and $C_1, \ldots, C_n$ are arbitrary constants. | The structure of the solution of (9.6). (For linear independence, see (10.12).) |
| 9.8 | The general solution of (9.5) is $$x_t = C_1 u_1(t) + \cdots + C_n u_n(t) + u_t^*$$ where $C_1 u_1(t) + \cdots + C_n u_n(t)$ is the general solution of (9.6) and $u_t^*$ is some solution of (9.5). | The structure of the solution of (9.5). |

| | | |
|---|---|---|
| 9.9 | $x_{t+2} + ax_{t+1} + bx_t = 0$ has the solution:<br>• For $\frac{1}{4}a^2 - b > 0$: $x_t = C_1 m_1^t + C_2 m_2^t$<br>• For $\frac{1}{4}a^2 - b = 0$: $x_t = (C_1 + C_2 t)(-a/2)^t$<br>• For $\frac{1}{4}a^2 - b < 0$: $x_t = Ar^t \cos(\theta t + \omega)$<br>where $m_{1,2} = -\frac{1}{2}a \pm \sqrt{\frac{1}{4}a^2 - b}$, $r = \sqrt{b}$,<br>$\cos\theta = -a/2\sqrt{b}$. | The solutions of a homogeneous linear second order difference equation with constant coefficients $a$ and $b$. $C_1$, $C_2$ and $\omega$ are arbitrary constants. |
| 9.10 | $x_{t+n} + a_1 x_{t+n-1} + \cdots + a_{n-1} x_{t+1} + a_n x_t = b(t)$ | The *general linear difference equation* of the $n$th order with constant coefficients. |
| 9.11 | $x_{t+n} + a_1 x_{t+n-1} + \cdots + a_{n-1} x_{t+1} + a_n x_t = 0$ | The *homogeneous* equation associated with (9.10). |
| 9.12 | $m^n + a_1 m^{n-1} + \cdots + a_{n-1} m + a_n = 0$ | The *characteristic equation* of (9.10) and (9.11). |
| 9.13 | To obtain $n$ linearly independent solutions of (9.11): Find all roots of (9.12).<br>• A real root $m_i$ with multiplicity 1 gives rise to a solution $Am_i^t$.<br>• A real root $m_j$ with multiplicity p gives rise to solutions $B_1 m_j^t, B_2 t m_j^t, \ldots, B_p t^{p-1} m_j^t$.<br>• A pair of complex roots $m_k = \alpha + i\beta$, $\overline{m}_k = \alpha - i\beta$ with multiplicity 1 gives rise to the solutions $C_1 r^t \cos\theta t$, $C_2 r^t \sin\theta t$, where $r = \sqrt{\alpha^2 + \beta^2}$, and $\theta \in [0, \pi]$ satisfies $\cos\theta = \alpha/r$.<br>• A pair of complex roots $m_e = \lambda + i\mu$, $\overline{m}_e = \lambda - i\mu$ with multiplicity $q$ gives rise to the solutions $D_1 u, D_2 v, D_3 tu, D_4 tv, \ldots, D_{2q-1} t^{q-1} u, D_{2q} t^{q-1} v$, with $u = s^t \cos\varphi t$, $v = s^t \sin\varphi t$, where $s = \sqrt{\lambda^2 + \mu^2}$, and $\varphi \in [0, \pi]$ satisfies $\cos\varphi = \lambda/s$. | A general method for finding $n$ linearly independent solutions to (9.11). |
| 9.14 | Equation (9.10) is *stable (globally asymptotically stable)* if the general solution $C_1 u_1(t) + \cdots + C_n u_n(t)$ of (9.11) approaches 0 as $t \to \infty$, for all values of $C_1, \ldots, C_n$. | Definition of stability for a linear equation with constant coefficients. |
| 9.15 | Equation (9.10) is stable $\Leftrightarrow$ All the characteristic roots of (9.10) have moduli less than 1. | Stability criterion for (9.10). |

| | | |
|---|---|---|
| 9.16 | $x_{t+1} + a_1 x_t = b(t)$ is stable $\iff$ $\|a_1\| < 1$ | Special case of (9.15). |
| 9.17 | $x_{t+2} + a_1 x_{t+1} + a_2 x_t = b(t)$ is stable $\iff$ $1 + a_1 + a_2 > 0$, $1 > a_1 - a_2$ and $1 > a_2$ | Special case of (9.15). |

$$\begin{vmatrix} 1 & a_n \\ \hline a_n & 1 \end{vmatrix} > 0, \quad \begin{vmatrix} 1 & 0 & a_n & a_{n-1} \\ a_1 & 1 & 0 & a_n \\ \hline a_n & 0 & 1 & a_1 \\ a_{n-1} & a_n & 0 & 1 \end{vmatrix} > 0, \ldots,$$

9.18
$$\begin{vmatrix} 1 & 0 & \cdots & 0 & a_n & a_{n-1} & \cdots & a_1 \\ a_1 & 1 & \cdots & 0 & 0 & a_n & \cdots & a_2 \\ \vdots & \vdots & \ddots & \vdots & \vdots & \vdots & \ddots & \vdots \\ a_{n-1} & a_{n-2} & \cdots & 1 & 0 & 0 & \cdots & a_n \\ \hline a_n & 0 & \cdots & 0 & 1 & a_1 & \cdots & a_{n-1} \\ a_{n-1} & a_n & \cdots & 0 & 0 & 1 & \cdots & a_{n-2} \\ \vdots & \vdots & \ddots & \vdots & \vdots & \vdots & \ddots & \vdots \\ a_1 & a_2 & \cdots & a_n & 0 & 0 & \cdots & 1 \end{vmatrix} > 0$$

A necessary and sufficient condition for all the roots of (9.12) to have moduli less than 1. (*Schur's theorem.*)

9.19
$$x_1(t+1) = a_{11}(t)x_1(t) + \cdots + a_{1n}(t)x_n(t) + b_1(t)$$
$$\cdots\cdots\cdots\cdots\cdots\cdots\cdots\cdots\cdots\cdots\cdots\cdots\cdots$$
$$x_n(t+1) = a_{n1}(t)x_1(t) + \cdots + a_{nn}(t)x_n(t) + b_n(t)$$

*Linear system of difference equations.*

9.20 $\mathbf{x}(t+1) = \mathbf{A}(t)\mathbf{x}(t) + \mathbf{b}(t)$, $t = 0, 1, \ldots$

Matrix form of (9.19). $\mathbf{x}(t)$ and $\mathbf{b}(t)$ are $n \times 1$, $\mathbf{A}(t)$ is $n \times n$.

9.21 $\mathbf{x}(t) = \mathbf{A}^t \mathbf{x}(0) + (\mathbf{A}^{t-1} + \mathbf{A}^{t-2} + \cdots + \mathbf{A} + \mathbf{I})\mathbf{b}$

The solution of (9.20) for $\mathbf{A}(t) = \mathbf{A}$, $\mathbf{b}(t) = \mathbf{b}$.

9.22 $\mathbf{x}(t+1) = \mathbf{A}\mathbf{x}(t) \iff \mathbf{x}(t) = \mathbf{A}^t \mathbf{x}(0)$

A special case of (9.21) where $\mathbf{b} = \mathbf{0}$, and with $\mathbf{A}^0 = \mathbf{I}$.

9.23 The difference equation (9.20) with $\mathbf{A}(t) = \mathbf{A}$ is called stable if $\mathbf{A}^t \mathbf{x}(0)$ converges to the zero vector for every choice of the vector $\mathbf{x}(0)$.

Definition of *stability* of a linear system.

9.24 The difference equation (9.20) with $\mathbf{A}(t) = \mathbf{A}$ is stable if and only if all the characteristic roots of $\mathbf{A}$ have moduli less than 1.

Characterization of stability of a linear system. (Sufficient conditions for the characteristic roots to have moduli less than 1 in terms of matrix norms follows from (19.27).)

9.25 | If all characteristic roots of $\mathbf{A} = (a_{ij})_{n \times n}$ have moduli less than 1, then every solution $\mathbf{x}(t)$ of
$$\mathbf{x}(t+1) = \mathbf{A}\mathbf{x}(t) + \mathbf{b}, \quad t = 0, 1, \ldots$$
converges to the vector $(\mathbf{I} - \mathbf{A})^{-1}\mathbf{b}$. | The solution of an important equation. (See (21.23).)

## References

The formulas and results are found in e.g. Hildebrand (1968) and Goldberg (1961).

# Chapter 10

# Differential equations

## First-order equations

10.1    $\dot{x}(t) = f(t)$    $\Longleftrightarrow$    $x(t) = x(t_0) + \int_{t_0}^{t} f(\tau)\, d\tau$
  
The simplest differential equation and its solution. $f(t)$ is a given continuous function and $x(t)$ is the unknown function.

10.2    $\dfrac{dx}{dt} = f(t)g(x)$    $\Longleftrightarrow$    $\int \dfrac{dx}{g(x)} = \int f(t)\, dt$

Separable differential equation. $(g(x) \neq 0.)$ If $g(a) = 0$, $x(t) \equiv a$ is a solution.

10.3    $\dot{x} = g(x/t)$ and $z = x/t \Rightarrow t\dfrac{dz}{dt} = g(z) - z$

The *projective* (or *homogeneous*) differential equation. The substitution $z = x/t$ leads to a separable equation for $z$.

10.4    $\dfrac{dx}{dt} = B(x-a)(x-b)$    $\Leftrightarrow$    $x = a + \dfrac{b-a}{1 - Ce^{B(b-a)t}}$

$a \neq b$. $a = 0$ gives the *logistic* equation.

10.5    $\dot{x} + ax = b$    $\Longleftrightarrow$    $x = Ce^{-at} + \dfrac{b}{a}$

Linear first-order differential equation with constant coefficients $a$ and $b$. $C$ is an arbitrary constant.

10.6    $\dot{x} + a(t)x = b(t)$ $\Longleftrightarrow$
  
$x = e^{-\int a(t)\, dt} \left( C + \int e^{\int a(t)\, dt} b(t)\, dt \right)$

General linear first-order differential equation. $a(t)$ and $b(t)$ are given. $C$ is the constant of integration.

10.7 $\dot{x} + a(t)x = b(t) \iff$
$$x(t) = x_0 e^{-\int_{t_0}^{t} a(\xi)\,d\xi} + \int_{t_0}^{t} b(\tau) e^{-\int_{\tau}^{t} a(\xi)\,d\xi}\,d\tau$$

General linear first-order differential equation. Solution with given initial condition $x(t_0) = x_0$.

10.8 $\dot{x} = Q(t)x + R(t)x^n$ has the solution
$$x(t) = e^{\frac{P(t)}{1-n}}\left[C + (1-n)\int R(t) e^{-P(t)}\,dt\right]^{\frac{1}{1-n}}$$
where $P(t) = (1-n)\int Q(t)\,dt$

Bernoulli's equation and its solution. $(n \neq 1)$

10.9 $\dot{x} = P(t) + Q(t)x + R(t)x^2$

Riccati's equation. Not analytically solvable in general. The substitution $x = u + 1/z$ works if we know a special solution $u = u(t)$.

## Higher order equations

10.10 $$\frac{d^n x}{dt^n} + a_1(t)\frac{d^{n-1}x}{dt^{n-1}} + \cdots + a_{n-1}(t)\frac{dx}{dt} + a_n(t)x = f(t)$$

The general linear $n$th-order differential equation.

10.11 $$\frac{d^n x}{dt^n} + a_1(t)\frac{d^{n-1}x}{dt^{n-1}} + \cdots + a_{n-1}(t)\frac{dx}{dt} + a_n(t)x = 0$$

The *homogeneous equation* associated with (10.10).

10.12 The functions $u_1(t),\ldots,u_m(t)$ are *linearly independent* if the equation
$$C_1 u_1(t) + \cdots + C_m u_m(t) = 0$$
holds for all $t$ only if $C_1,\ldots,C_m$ are all 0.

Definition of linear independence.

10.13 The general solution of (10.11) is
$$x(t) = C_1 u_1(t) + \cdots + C_n u_n(t)$$
where $u_1(t),\ldots,u_n(t)$ are $n$ linearly independent solutions of (10.11) and $C_1,\ldots,C_n$ are arbitrary constants.

The structure of the solution of (10.11).

| | | |
|---|---|---|
| 10.14 | The general solution of (10.10) is<br>$$x(t) = C_1 u_1(t) + \cdots + C_n u_n(t) + u^*(t)$$<br>where $C_1 u_1(t) + \cdots + C_n u_n(t)$ is the general solution of the homogeneous equation (10.11), and $u^*(t)$ is some solution of equation (10.10). | The structure of the solution of (10.10). |
| 10.15 | $\ddot{x} + a\dot{x} + bx = 0$ has the solution:<br>• For $\frac{1}{4}a^2 - b > 0$: $x = C_1 e^{r_1 t} + C_2 e^{r_2 t}$<br>• For $\frac{1}{4}a^2 - b = 0$: $x = (C_1 + C_2 t)e^{-at/2}$<br>• For $\frac{1}{4}a^2 - b < 0$: $x = Ae^{\alpha t}\cos(\beta t + \omega)$<br>where $r_{1,2} = -\frac{1}{2}a \pm \sqrt{\frac{1}{4}a^2 - b}$, $\alpha = -\frac{1}{2}a$,<br>$\beta = \sqrt{b - \frac{1}{4}a^2}$ | The solutions of a homogeneous second-order linear differential equation with constant coefficients $a$ and $b$. $C_1$, $C_2$, $A$, and $\omega$ are constants. |
| 10.16 | $\ddot{x} + a\dot{x} + bx = f(t)$, $b \neq 0$, has a special solution $u^* = u^*(t)$:<br>• $f(t) = A$: $u^* = A/b$<br>• $f(t) = At + B$: $u^* = \frac{A}{b}t + \frac{bB - aA}{b^2}$<br>• $f(t) = At^2 + Bt + C$:<br>$u^* = \frac{At^2}{b} + \frac{(bB - 2aA)t}{b^2} + \frac{Cb^2 - (2A + aB)b + 2a^2 A}{b^3}$<br>• $f(t) = pe^{qt}$: $u^* = pe^{qt}/(q^2 + aq + b)$<br>if $q^2 + aq + b \neq 0$. | Special solutions of non-homogeneous second-order linear differential equations with constant coefficients $a$ and $b$. |
| 10.17 | $\dfrac{d^n x}{dt^n} + a_1 \dfrac{d^{n-1} x}{dt^{n-1}} + \cdots + a_{n-1}\dfrac{dx}{dt} + a_n x = f(t)$ | The general $n$th-order linear differential equation with constant coefficients. |
| 10.18 | $\dfrac{d^n x}{dt^n} + a_1 \dfrac{d^{n-1} x}{dt^{n-1}} + \cdots + a_{n-1}\dfrac{dx}{dt} + a_n x = 0$ | The *homogeneous equation* associated with (10.17). |
| 10.19 | $r^n + a_1 r^{n-1} + \cdots + a_{n-1} r + a_n = 0$ | The *characteristic equation* associated with (10.17) and (10.18). |
| 10.20 | $x = x(t) = C_1 e^{r_1 t} + C_2 e^{r_2 t} + \cdots + C_n e^{r_n t}$ | The general solution of (10.18) if the roots $r_1, \ldots, r_n$ of (10.19) are all real and different. |

**10.21** To obtain $n$ linearly independent solutions of (10.18): Find all roots of (10.19).
- A real root $r_i$ with multiplicity 1 gives rise to a solution $Ae^{r_i t}$.
- A real root $r_j$ with multiplicity p gives rise to solutions $B_1 e^{r_j t}, B_2 t e^{r_j t}, \ldots, B_p t^{p-1} e^{r_j t}$.
- A pair of complex roots $r_k = \alpha + i\beta$, $\bar{r}_k = \alpha - i\beta$ with multiplicity 1 gives rise to the solutions $C_1 e^{\alpha t} \cos \beta t$, $C_2 e^{\alpha t} \sin \beta t$.
- A pair of complex roots $r_e = \lambda + i\mu$, $\bar{r}_e = \lambda - i\mu$ with multiplicity $q$ gives rise to the solutions: $D_1 u, D_2 v, D_3 tu, D_4 tv, \ldots, D_{2q-1} t^{q-1} u, D_{2q} t^{q-1} v$, with $u = e^{\lambda t} \cos \mu t$, $v = e^{\lambda t} \sin \mu t$.

*General method for finding $n$ linearly independent solutions of (10.18).*

**10.22** Method for obtaining a special solution of (10.17) if $u_1, \ldots, u_n$ are $n$ linearly independent solutions of (10.18): Solve

$$\dot{C}_1(t) u_1 + \cdots + \dot{C}_n(t) u_n = 0$$
$$\dot{C}_1(t) \dot{u}_1 + \cdots + \dot{C}_n(t) \dot{u}_n = 0$$
$$\cdots\cdots\cdots\cdots\cdots\cdots\cdots\cdots\cdots\cdots\cdots\cdots$$
$$\dot{C}_1(t) u_1^{(n-2)} + \cdots + \dot{C}_n(t) u_n^{(n-2)} = 0$$
$$\dot{C}_1(t) u_1^{(n-1)} + \cdots + \dot{C}_n(t) u_n^{(n-1)} = b(t)$$

for $\dot{C}_1(t), \ldots, \dot{C}_n(t)$. Integrate to find $C_1(t), \ldots, C_n(t)$. Then a special solution of (10.17) is: $u^*(t) = C_1(t) u_1 + \cdots + C_n(t) u_n$.

*General method for finding a special solution of (10.17), called variation of parameters. ($u_j^{(i)}$ denotes the derivative $d^i u_j / dt^i$.)*

**10.23** Equation (10.17) is *stable (globally asymptotically stable)* if the general solution $C_1 u_1(t) + \cdots + C_n u_n(t)$ of (10.18) approaches 0 as $t \to \infty$, for all values of $C_1, \ldots, C_n$.

*Definition of stability for a linear equation with constant coefficients.*

**10.24** Equation (10.17) is stable $\Leftrightarrow$ All the characteristic roots of (10.17) have negative real parts.

*Stability criterion for (10.17).*

**10.25**
$\dot{x} + ax = f(t)$ is stable $\iff a > 0$

$\ddot{x} + a\dot{x} + bx = f(t)$ is stable $\iff \begin{cases} a > 0 \\ b > 0 \end{cases}$

*Special cases of (10.24).*

**10.26** (10.17) is stable $\Rightarrow a_i > 0$ for all $i = 1, \ldots, n$

*Necessary condition for the stability of (10.17).*

10.27 $$\mathbf{A} = \begin{pmatrix} a_1 & a_3 & a_5 & \cdots & 0 & 0 \\ a_0 & a_2 & a_4 & \cdots & 0 & 0 \\ 0 & a_1 & a_3 & \cdots & 0 & 0 \\ \vdots & \vdots & \vdots & \ddots & \vdots & \vdots \\ 0 & 0 & 0 & \cdots & a_{n-1} & 0 \\ 0 & 0 & 0 & \cdots & a_{n-2} & a_n \end{pmatrix}$$

A matrix associated with the numbers $a_0, a_1, \ldots, a_n$. (The $k$th column of this matrix is $\ldots a_{k+1}\, a_k\, a_{k-1} \ldots$ where the element $a_k$ is on the main diagonal. An elements $a_{k+j}$ with $k+j$ negative or greater than $n$, is set equal to 0.)

10.28 (10.17) is stable $\iff$ $\begin{cases} \text{All the leading principal} \\ \text{minors of } \mathbf{A} \text{ in (10.27)} \\ \text{are positive } (a_0 = 1). \end{cases}$

*Routh-Hurwitz's* stability conditions.

## Systems of differential equations

10.29
$$\dot{x}_1 = f_1(x_1, \ldots, x_n)$$
$$\cdots\cdots\cdots\cdots\cdots$$
$$\dot{x}_n = f_n(x_1, \ldots, x_n)$$

An *autonomous* system of first-order differential equations.

10.30 $\mathbf{a} = (a_1, \ldots, a_n)$ is an *equilibrium point* for the system (10.29) if $f_i(\mathbf{a}) = 0$, $i = 1, \ldots, n$.

Definition of an equilibrium for (10.29).

10.31 The equilibrium point $\mathbf{a}$ for (10.29) is *(locally) stable* if all solutions that come close to $\mathbf{a}$ stay close to $\mathbf{a}$: For every $\varepsilon > 0$ there is a $\delta > 0$ such that if $||\mathbf{x} - \mathbf{a}|| < \delta$, then there exists a solution $\varphi(t)$ of (10.29), defined for $t \geq 0$, with $\varphi(0) = \mathbf{x}$, which satisfies

$$||\varphi(t) - \mathbf{a}|| < \varepsilon \quad \text{for all } t > 0$$

If $\mathbf{a}$ is stable and there exists a $\delta_0 > 0$ such that

$$||\mathbf{x} - \mathbf{a}|| < \delta_0 \implies \lim_{t \to \infty} ||\varphi(t) - \mathbf{a}|| = 0$$

then $\mathbf{a}$ is (locally) asymptotically stable.
If $\mathbf{a}$ is not stable, it is called *unstable*.

Definition of (local) stability.

10.32 If every solution of (10.29), whatever its initial point, converges to a unique equilibrium point $\mathbf{a}$, then $\mathbf{a}$ is *globally asymptotically stable*.

Global asymptotic stability.

10.33　Let **a** be an equilibrium point for (10.29) and define

$$\mathbf{A} = \begin{pmatrix} \dfrac{\partial f_1(\mathbf{a})}{\partial x_1} & \cdots & \dfrac{\partial f_1(\mathbf{a})}{\partial x_n} \\ \vdots & \ddots & \vdots \\ \dfrac{\partial f_n(\mathbf{a})}{\partial x_1} & \cdots & \dfrac{\partial f_n(\mathbf{a})}{\partial x_n} \end{pmatrix}$$

If all the characteristic roots of **A** have negative real parts, then **a** is (locally) asymptotically stable.

If at least one characteristic root of **A** has a positive real part, then **a** is unstable.

*A Liapunov theorem.*

10.34　Let $(a, b)$ be an equilibrium point for the system $\dot{x} = f(x, y)$, $\dot{y} = g(x, y)$, and define

$$\mathbf{A} = \begin{pmatrix} \dfrac{\partial f(a,b)}{\partial x} & \dfrac{\partial f(a,b)}{\partial y} \\ \dfrac{\partial y(u,b)}{\partial x} & \dfrac{\partial g(a,b)}{\partial y} \end{pmatrix}$$

Then, if $\operatorname{tr}(\mathbf{A}) < 0$ and $\det(\mathbf{A}) > 0$, $(a,b)$ is locally asymptotically stable.

*A special case of (10.33). Stability in terms of the signs of the trace and the determinant of $\mathbf{A}$, valid if $n = 2$.*

10.35　Necessary and sufficient conditions for all the characteristic roots of a $3 \times 3$-matrix $B = (b_{ij})$ to have negative real parts is that the following inequalities hold:

- $\operatorname{tr}(B) < 0$, $\det(B) < 0$, and

- $\begin{vmatrix} b_{11}+b_{22} & b_{23} & -b_{13} \\ b_{32} & b_{11}+b_{33} & b_{12} \\ -b_{31} & b_{21} & b_{22}+b_{33} \end{vmatrix} < 0$

*The modified Routh-Hurwitz's stability conditions.*

10.36　$V(\mathbf{x}) = V(x_1, \ldots, x_n)$ is a Liapunov function for system (10.29) in an open set $\Omega$ containing an equilibrium point **a** if

- $V(\mathbf{x}) > 0$ for all $\mathbf{x} \in \Omega$, $\mathbf{x} \neq \mathbf{a}$, $V(\mathbf{a}) = 0$

- $\dot{V}(\mathbf{x}) = \sum_{i=1}^{n} \dfrac{\partial V(\mathbf{x})}{\partial x_i} \dfrac{dx_i}{dt} = \sum_{i=1}^{n} \dfrac{\partial V(\mathbf{x})}{\partial x_i} f_i(\mathbf{x}) \leq 0$

  for all $\mathbf{x} \in \Omega$, $\mathbf{x} \neq \mathbf{a}$

*Definition of a Liapunov function.*

10.37 Let **a** be an equilibrium point for (10.29) and suppose there exists a Liapunov function $V(\mathbf{x})$ for the system in an open set $\Omega$ containing **a**. Then **a** is a stable equilibrium point. If also
$$\dot{V}(\mathbf{x}) < 0 \text{ for all } \mathbf{x} \in \Omega, \mathbf{x} \neq \mathbf{a},$$
then **a** is locally asymptotically stable.

*Another Liapunov theorem.*

10.38 The *Lotka-Volterra model*
$$\dot{x} = kx - axy, \qquad \dot{y} = -hy + bxy$$
has an equilibrium $(h/b, k/a)$ which is stable (but not asymptotically stable).

*Possible interpretation: $x$ is the number of rabbits, $y$ is the number of foxes. ($a$, $b$, $h$, and $k$ are positive.)*

10.39 The modified Lotka-Volterra model
$$\dot{x} = kx - axy - \varepsilon x^2, \qquad \dot{y} = -hy + bxy - \delta y^2$$
has an asymptotically stable equilibrium
$$(x_0, y_0) = \left( \frac{ah + k\delta}{ab + \delta\varepsilon}, \frac{bk - h\varepsilon}{ab + \delta\varepsilon} \right)$$

*"The number of rabbits grows logistically in the absence of foxes." ($a$, $b$, $h$, $k$, $\delta$, and $\varepsilon$ are positive, $bk > h\varepsilon$.)*

10.40 Let $(a, b)$ be an equilibrium point for the system $\dot{x} = f(x, y)$, $\dot{y} = g(x, y)$, and define **A** as the matrix in (10.34). Then, if the characteristic roots of **A** are real and of opposite signs, there exist precisely two solutions $(x_1(t), y_1(t))$ and $(x_2(t), y_2(t))$ defined on an interval $[t_0, \infty)$ which converge to $(a, b)$. These solutions converge to $(a, b)$ from opposite directions, and both are tangent to the line through $(a, b)$ parallel to the characteristic vector corresponding to the negative characteristic root. Such an equilibrium is called a *saddle point*.

*A local saddle point theorem.*

10.41 Consider the initial value problem
(∗) $\dot{\mathbf{x}} = \mathbf{F}(t, \mathbf{x}), \quad \mathbf{x}(t_0) = \mathbf{x}^0$
where $\mathbf{F}(t, \mathbf{x})$ and $\mathbf{F}'_{\mathbf{x}}(t, \mathbf{x})$ are continuous over
$$\Gamma = \{ (t, \mathbf{x}) : |t - t_0| \leq a, \, \|\mathbf{x} - \mathbf{x}^0\| \leq b \}$$
Define
$$M = \max_{(t, \mathbf{x}) \in \Gamma} \|\mathbf{F}(t, \mathbf{x})\|, \quad r = \min(a, b/M)$$
Then (∗) has a unique solution $\mathbf{x}(t)$ on the open interval $(t_0 - r, t_0 + r)$, and $\|\mathbf{x}(t) - \mathbf{x}^0\| \leq b$ in this interval.

*A (local) existence and uniqueness theorem for differential equations. $\mathbf{x} = (x_1, \ldots, x_n)$, $\mathbf{x}^0 = (x_1^0, \ldots, x_n^0)$.*

10.42 Suppose that $\mathbf{F}(t,\mathbf{x})$ and $\mathbf{F}'_{\mathbf{x}}(t,\mathbf{x})$ are continuous for all $(t,\mathbf{x})$ and that there exist continuous functions $a(t)$ and $b(t)$ such that

(∗) $\|\mathbf{F}(t,\mathbf{x})\| \leq a(t)\|\mathbf{x}\| + b(t)$  for all $(t,\mathbf{x})$

or

(∗∗) $\mathbf{x}\mathbf{F}(t,\mathbf{x}) \leq a(t)\|\mathbf{x}\|^2 + b(t)$  for all $(t,\mathbf{x})$

Then, given any point $(t_0, \mathbf{x}^0)$, there exists a solution $\mathbf{x}(t)$ of $\dot{\mathbf{x}} = \mathbf{F}(t,\mathbf{x})$ with $\mathbf{x}(t_0) = \mathbf{x}^0$, defined on $(-\infty, \infty)$.

The inequality (∗) is satisfied, in particular, if for all $(t,\mathbf{x})$,

(∗∗∗) $\|\mathbf{F}'_{\mathbf{x}}(t,\mathbf{x})\| \leq c(t)$  for a continuous $c(t)$

In this case the solution is unique.

A *global existence and uniqueness theorem* for differential equations. $\mathbf{x} = (x_1, \ldots, x_n)$, $\mathbf{x}^0 = (x_1^0, \ldots, x_n^0)$. In (∗∗∗) any matrix norm for $\mathbf{F}'_{\mathbf{x}}(t,\mathbf{x})$ can be used. (For matrix norms see (19.25).)

## Partial differential equations

Method for solving

(∗) $P(x,y,z)\dfrac{\partial z}{\partial x} + Q(x,y,z)\dfrac{\partial z}{\partial y} = R(x,y,z)$

- Find the solutions of the system
$$\frac{dy}{dx} = \frac{Q}{P}, \quad \frac{dz}{dx} = \frac{R}{P}$$
where $x$ is the free variable. If the solutions are
10.43  $y = \varphi_1(x, C_1, C_2)$, $z = \varphi_2(x, C_1, C_2)$, solve for $C_1$ and $C_2$ to obtain $u(x,y,z) = C_1$, $v(x,y,z) = C_2$.

- The general solution of (∗) is then

$$\Phi\bigl(u(x,y,z), v(x,y,z)\bigr) = 0$$

where $\Phi$ is an arbitrary $C^1$-function of two variables, and at least one of the functions $u$ and $v$ contains $z$.

The *linear first-order partial differential equation* and a solution method.

## References

Braun (1983) is a good reference for most of the results in this chapter. See also Pontryagin (1962). For (10.27)–(10.28) see Takayama (1985). Beavis and Dobbs (1990) have most of the qualitative results and also economic applications. For (10.43) see Sneddon (1957).

# Chapter 11

# Topological concepts in Euclidean space

| | | |
|---|---|---|
| 11.1 | $B(\mathbf{a}; r) = \{\mathbf{x} : \|\mathbf{x} - \mathbf{a}\| < r\}$ | Definition of an *open n-ball* of radius $r$ and center $\mathbf{a}$ in $R^n$. |
| 11.2 | $\mathbf{a} \in S \subset R^n$ is an *interior point* of $S$ if there is an open $n$-ball with center at $\mathbf{a}$, all of whose points belong to $S$. | Definition of an interior point in $R^n$. |
| 11.3 | A set $S$ in $R^n$ is called *open* if all its points are interior points. | Definition of an open set in $R^n$. |
| 11.4 | A set $S$ in $R^n$ is called *closed* if $R^n \setminus S$ is open. | Definition of a closed set. $R^n \setminus S = \{\mathbf{x} \in R^n : \mathbf{x} \notin S\}$ |
| 11.5 | A set $S$ in $R^n$ is *bounded* if there exists a number $M$ such that $\|\mathbf{x}\| \leq M$ for all $\mathbf{x} \in S$ | Definition of a bounded set. |
| 11.6 | A sequence $\{\mathbf{x}_k\}$ in $R^n$ *converges* to $\mathbf{x}$ if for every $\varepsilon > 0$ there exists an integer $N$ such that $\|\mathbf{x}_k - \mathbf{x}\| < \varepsilon$ for all $k \geq N$. | Convergence of a sequence in $R^n$. |
| 11.7 | A sequence $\{\mathbf{x}_k\}$ in $R^n$ is a *Cauchy sequence* if for every $\varepsilon > 0$ there exists an integer $N$ such that $\|\mathbf{x}_j - \mathbf{x}_k\| < \varepsilon$ for all $j, k \geq N$. | Definition of a Cauchy sequence. |
| 11.8 | A sequence $\{\mathbf{x}_k\}$ in $R^n$ converges if and only if it is a Cauchy sequence. | *Cauchy's criterion.* |
| 11.9 | A set $S$ in $R^n$ is *closed* if and only if the limit $\mathbf{x} = \lim_k \mathbf{x}_k$ of each convergent sequence $\{\mathbf{x}_k\}$ of points in $S$ also lies in $S$. | Characterization of a closed set. |
| 11.10 | Let $\{\mathbf{x}_k\}$ be a sequence in $R^n$, and let $k_1 < k_2 < k_3 < \cdots$ be an increasing sequence of integers. Then $\{\mathbf{x}_{k_j}\}$, $j = 1, 2, \ldots$, is called a *subsequence* of $\{\mathbf{x}_k\}$. | Definition of a subsequence. |

| | | |
|---|---|---|
| 11.11 | A set $S$ in $R^n$ is *compact* (closed and bounded) if and only if every sequence of points in $S$ has a subsequence which converges to a point in $S$. | Characterization of a compact set. |
| 11.12 | $f : M \subset R^n \to R$ is *continuous* at $\mathbf{a} \in M$ if for each $\varepsilon > 0$ there exists a $\delta > 0$ such that $$|f(\mathbf{x}) - f(\mathbf{a})| < \varepsilon$$ for every $\mathbf{x} \in M$ with $||\mathbf{x} - \mathbf{a}|| < \delta$. | Definition of a continuous function of $n$ variables. |
| 11.13 | The transformation $\mathbf{f} = (f_1, \ldots, f_m) : M \subset R^n \to R^m$ is called *continuous* at $\mathbf{a}$ if each $f_i$ is continuous at $\mathbf{a}$ according to definition (11.12). | Definition of a continuous (vector) function from $R^n$ to $R^m$. |
| 11.14 | The transformation $\mathbf{f} = (f_1, \ldots, f_m) : M \subset R^n \to R^m$ is continuous at $\mathbf{a} \in M$ if and only if $\mathbf{f}(\mathbf{x}_k) \to \mathbf{f}(\mathbf{a})$ for every sequence $\{\mathbf{x}_k\}$ which lies in $M$ and converges to $\mathbf{a}$. | Characterization of a continuous (vector) function. |
| 11.15 | A transformation $\mathbf{f} : R^n \to R^m$ is continuous at each point $\mathbf{x}$ in $R^n$ if and only if $\mathbf{f}^{-1}(T)$ is open (closed) for every open (closed) set $T$ in $R^m$. | Characterization of a continuous (vector) function. |
| 11.16 | If $\mathbf{f}$ is a continuous transformation of $R^n$ into $R^m$ and $M$ is a compact set in $R^n$, then $\mathbf{f}(M)$ is compact. | Continuous transformations map compact sets onto compact sets. |
| 11.17 | If $f : S \to R$ is continuous on a closed, bounded subset $S$ of $R^n$, then there exist maximum and minimum points for $f$ in $S$, i.e. points $\mathbf{c}, \mathbf{d} \in S$ such that $$f(\mathbf{c}) \geq f(\mathbf{x}) \geq f(\mathbf{d}) \quad \text{for all } \mathbf{x} \in S$$ | The *extreme value theorem* (or *Weierstrass's theorem*). |
| 11.18 | For each natural number $n$ let $\mathbf{f}_n$ be a function defined on a set $S \subset R^n$ and with range in $R^m$. The sequence $\{\mathbf{f}_n\}$ is said to *converge pointwise* to a function $\mathbf{f}$ on a subset $S_0$ of $S$, if for each $\mathbf{x} \in S_0$, the sequence $\{\mathbf{f}_n(\mathbf{x})\}$ converges in $R^m$ to $\mathbf{f}(\mathbf{x})$. | Definition of (pointwise) convergence of a sequence of functions. |
| 11.19 | The sequence $\{\mathbf{f}_n\}$, where $\mathbf{f}_n$ is defined on a set $S \subset R^n$ and with range in $R^m$, is said to *converge uniformly* to a function $\mathbf{f}$ on a subset $S_0$ of $S$, if for each $\varepsilon > 0$ there is a natural number $N(\varepsilon)$ (depending on $\varepsilon$ but NOT on $\mathbf{x} \in S_0$) such that for all $n \geq N(\varepsilon)$ and $\mathbf{x} \in S_0$, then $$||\mathbf{f}_n(\mathbf{x}) - \mathbf{f}(\mathbf{x})|| < \varepsilon$$ | Definition of uniform convergence of a sequence of functions. |

11.20 A *correspondence* $F$ from a set $A$ to a set $B$ is a rule which to each $x \in A$ associates a non-empty subset $F(x)$ of $B$.

*Definition of a correspondence.*

11.21 The correspondence $\mathbf{F}: X \subset R^n \to R^m$ is said to have a *closed graph* if for every pair of sequences $\{\mathbf{x}_k\}$ and $\{\mathbf{y}_k\}$, where $\mathbf{x}_k \in X$ and $\mathbf{y}_k \in \mathbf{F}(\mathbf{x}_k)$, if $\mathbf{x}_k \to \mathbf{x}$ with $\mathbf{x} \in X$ and $\mathbf{y}_k \to \mathbf{y}$, then $\mathbf{y} \in \mathbf{F}(\mathbf{x})$.

*Definition of a closed graph of a correspondence.*

11.22 The correspondence $\mathbf{F}: X \subset R^n \to R^m$ is said to be *lower hemicontinuous* at $\mathbf{x}^0$ if for every $\varepsilon > 0$ there exists a $\delta > 0$ such that if $\mathbf{x} \in X$, $\|\mathbf{x} - \mathbf{x}_0\| < \delta$, and $\mathbf{y}_0 \in \mathbf{F}(\mathbf{x}_0)$, then $\mathbf{F}(\mathbf{x}) \cap B(\mathbf{y}_0; \varepsilon) \neq \emptyset$.

*Definition of lower hemicontinuity of a correspondence.*

11.23 The correspondence $\mathbf{F}: X \subset R^n \to R^m$ is said to be *upper hemicontinuous* at $\mathbf{x}^0$ if for every open set $U$ containing $\mathbf{F}(\mathbf{x}_0)$ there exists a $\delta > 0$ such that if $\|\mathbf{x} - \mathbf{x}_0\| < \delta$ and $\mathbf{x} \in X$, then $\mathbf{F}(\mathbf{x}) \subset U$.

*Definition of upper hemicontinuity of a correspondence.*

11.24 Let $\mathbf{F}: X \subset R^n \to K \subset R^m$ be a correspondence where $K$ is compact. Suppose that for every $\mathbf{x} \in X$ the set $\mathbf{F}(\mathbf{x})$ is a closed subset of $K$. Then $\mathbf{F}$ has a closed graph if and only if $\mathbf{F}$ is upper hemicontinuous.

*An interesting result.*

## Infimum and supremum

11.25
$\inf A$ = The largest number less than or equal to all numbers in $A$.

$\sup A$ = The smallest number greater than or equal to all numbers in $A$.

*Definition of infimum and supremum of a set $A$ of real numbers.*

11.26
$$\inf_{\mathbf{x} \in B} f(\mathbf{x}) = \inf\{f(\mathbf{x}) : \mathbf{x} \in B\}$$

$$\sup_{\mathbf{x} \in B} f(\mathbf{x}) = \sup\{f(\mathbf{x}) : \mathbf{x} \in B\}$$

*Definition of infimum and supremum of a real valued function defined on a set $B$ in $R^n$.*

11.27 $\inf_{\mathbf{x} \in B}(f(\mathbf{x}) + g(\mathbf{x})) \geq \inf_{\mathbf{x} \in B} f(\mathbf{x}) + \inf_{\mathbf{x} \in B} g(\mathbf{x})$

*Result about sup and inf.*

| | | |
|---|---|---|
| 11.28 | $\inf_{\mathbf{x}\in B}(\lambda f(\mathbf{x})) = \lambda \inf_{\mathbf{x}\in B} f(\mathbf{x})$ ($\lambda \geq 0$ a real number) | Results about sup and inf. |
| 11.29 | $\sup_{\mathbf{x}\in B}(f(\mathbf{x}) + g(\mathbf{x})) \leq \sup_{\mathbf{x}\in B} f(\mathbf{x}) + \sup_{\mathbf{x}\in B} g(\mathbf{x})$ | |
| 11.30 | $\sup_{\mathbf{x}\in B}(\lambda f(\mathbf{x})) = \lambda \sup_{\mathbf{x}\in B} f(\mathbf{x})$ ($\lambda \geq 0$ a real number) | |
| 11.31 | $\sup_{\mathbf{x}\in B}(-f(\mathbf{x})) = -\inf_{\mathbf{x}\in B} f(\mathbf{x})$ | |
| 11.32 | $\inf_{\mathbf{x}\in B}(-f(\mathbf{x})) = -\sup_{\mathbf{x}\in B} f(\mathbf{x})$ | |
| 11.33 | $\sup_{(\mathbf{x},\mathbf{y})\in A\times B} f(\mathbf{x},\mathbf{y}) = \sup_{\mathbf{x}\in A}(\sup_{\mathbf{y}\in B} f(\mathbf{x},\mathbf{y}))$ | $A \times B = \{(\mathbf{x},\mathbf{y}) : \mathbf{x} \in A$ and $\mathbf{y} \in B\}$ |
| 11.34 | $\underline{\lim}_{\mathbf{x}\to\mathbf{x}^0} f(\mathbf{x}) = \lim_{r\to 0}(\inf\{f(\mathbf{x}) : \mathbf{x} \in B(\mathbf{x}^0; r) \cap M\})$ | Definition of $\underline{\lim}$ = lim inf. $f$ is defined on $M \subset R^n$ and $\mathbf{x}^0$ is in the closure $\overline{M}$ of $M$. ($\overline{M}$ is the smallest closed subset of $R^n$ containing $M$.) |
| 11.35 | $\overline{\lim}_{\mathbf{x}\to\mathbf{x}^0} f(\mathbf{x}) = \lim_{r\to 0}(\sup\{f(\mathbf{x}) : \mathbf{x} \in B(\mathbf{x}^0, r) \cap M\})$ | Definition of $\overline{\lim}$ = lim sup. $f$ is defined on $M \subset R^n$ and $\mathbf{x}^0$ is in the closure $\overline{M}$ of $M$. |
| 11.36 | $\underline{\lim}(f+g) \geq \underline{\lim} f + \underline{\lim} g$ | Results on lim inf and lim sup. |
| 11.37 | $\overline{\lim}(f+g) \leq \overline{\lim} f + \overline{\lim} g$ | |
| 11.38 | $\underline{\lim} f \leq \overline{\lim} f$ | |
| 11.39 | $\underline{\lim} f = -\overline{\lim}(-f)$, $\overline{\lim} f = -\underline{\lim}(-f)$ | |

## References

Bartle (1976) and Rudin (1982) are good references for standard topological results. For correspondences and their properties, see Hildenbrand and Kirman (1976) or Hildenbrand (1974).

# Chapter 12

# Convexity

12.1 A set $S$ in $R^n$ is *convex* if
$\mathbf{x}, \mathbf{y} \in S$ and $\lambda \in [0,1] \Rightarrow \lambda \mathbf{x} + (1-\lambda)\mathbf{y} \in S$
— Definition of a convex set.

12.2 If $S$ and $T$ are convex sets in $R^n$, then
- $S \cap T = \{\mathbf{x} : \mathbf{x} \in S \text{ and } \mathbf{x} \in T\}$ is convex.
- $aS + bT = \{a\mathbf{s} + b\mathbf{t} : \mathbf{s} \in S, \mathbf{t} \in T\}$ is convex.

— Properties of convex sets. ($a$ and $b$ are real numbers.)

12.3 Let $S$ and $T$ be convex sets in $R^n$ with no points in common. Then $S$ and $T$ can be separated by a hyperplane: There exists a vector $(a_1, \ldots, a_n) \neq \mathbf{0}$ such that for all $(x_1, \ldots, x_n) \in S$ and all $(y_1, \ldots, y_n) \in T$,
$$a_1 x_1 + \cdots + a_n x_n \leq a_1 y_1 + \cdots + a_n y_n$$

— Minkowski's separation theorem. The hyperplane in question is $\{\mathbf{x} : \mathbf{a} \cdot \mathbf{x} = A\}$, where $A = \sup\{\mathbf{a} \cdot \mathbf{x} : \mathbf{x} \in S\}$.

12.4 Let $S$ be a convex set in $R^n$ with interior points and let $T$ be a convex set in $R^n$ such that no point in $S \cap T$ (if there is any) is an interior point in $S$. Then $S$ and $T$ can be separated by a hyperplane, i.e. there exists a vector $\mathbf{a} \neq \mathbf{0}$ such that
$$\mathbf{a} \cdot \mathbf{x} \leq \mathbf{a} \cdot \mathbf{y} \quad \text{for all } \mathbf{x} \in S \text{ and all } \mathbf{y} \in T$$

— A general separation theorem in $R^n$.

12.5 $f(\mathbf{x}) = f(x_1, \ldots, x_n)$ defined on the convex set $S$ in $R^n$ is *concave* on $S$ if
$$f(\lambda \mathbf{x} + (1-\lambda)\mathbf{x}^0) \geq \lambda f(\mathbf{x}) + (1-\lambda) f(\mathbf{x}^0)$$
for all $\mathbf{x}, \mathbf{x}^0 \in S$ and all $\lambda \in (0,1)$.

— Definition of a concave function.

12.6 $f(\mathbf{x})$ is *strictly concave* if $f(\mathbf{x})$ is concave and the inequality $\geq$ in (12.5) is strict for $\mathbf{x} \neq \mathbf{x}^0$.

— Definition of a strictly concave function.

12.7 $f(\mathbf{x}) = f(x_1, \ldots, x_n)$ defined on the convex set $S$ in $R^n$ is *convex* on $S$ if
$$f(\lambda \mathbf{x} + (1-\lambda)\mathbf{x}^0) \leq \lambda f(\mathbf{x}) + (1-\lambda) f(\mathbf{x}^0)$$
for all $\mathbf{x}, \mathbf{x}^0 \in S$ and all $\lambda \in (0,1)$.

— Definition of a convex function.

| | | |
|---|---|---|
| 12.8 | $f(\mathbf{x})$ is *strictly convex* if $f(\mathbf{x})$ is convex and the inequality $\leq$ in (12.7) is strict for $\mathbf{x} \neq \mathbf{x}^0$. | Definition of a strictly convex function. |
| 12.9 | If $f(\mathbf{x})$, defined on the convex set $S$ in $R^n$, is concave (convex), then $f(\mathbf{x})$ is continuous at each interior point of $S$. | On continuity of concave (convex) functions. |
| 12.10 | If $f(\mathbf{x})$ and $g(\mathbf{x})$ are concave (convex) and $a$ and $b$ are nonnegative numbers, then $af(\mathbf{x}) + bg(\mathbf{x})$ is concave (convex). | Properties of concave and convex functions. |
| 12.11 | If $f(\mathbf{x})$ is concave and $F(u)$ is concave and increasing, then $U(\mathbf{x}) = F(f(\mathbf{x}))$ is concave. | |
| 12.12 | If $f(\mathbf{x}) = \mathbf{a} \cdot \mathbf{x} + b$ and $F(u)$ is concave, then $U(\mathbf{x}) = F(f(\mathbf{x}))$ is concave. | |
| 12.13 | If $f(\mathbf{x})$ is convex and $F(u)$ is convex and increasing, then $U(\mathbf{x}) = F(f(\mathbf{x}))$ is convex. | |
| 12.14 | If $f(\mathbf{x}) = \mathbf{a} \cdot \mathbf{x} + b$ and $F(u)$ is convex, then $U(\mathbf{x}) - F(f(\mathbf{x}))$ is convex. | |
| 12.15 | $f(\mathbf{x})$ is concave on an open, convex subset $S$ in $R^n$ if and only if for all $\mathbf{x}, \mathbf{x}^0 \in S$ $$f(\mathbf{x}) - f(\mathbf{x}^0) \leq \sum_{i=1}^{n} \frac{\partial f(\mathbf{x}^0)}{\partial x_i}(x_i - x_i^0)$$ | Characterization of concavity for $C^1$-functions. |
| 12.16 | $f(\mathbf{x})$ is strictly concave on an open, convex subset $S$ in $R^n$ if and only if the inequality in (12.15) is strict for $\mathbf{x} \neq \mathbf{x}^0$. | Strict concavity for $C^1$-functions. |
| 12.17 | $f(\mathbf{x})$ is convex on an open, convex subset $S$ in $R^n$ if and only if for all $\mathbf{x}, \mathbf{x}^0 \in S$ $$f(\mathbf{x}) - f(\mathbf{x}^0) \geq \sum_{i=1}^{n} \frac{\partial f(\mathbf{x}^0)}{\partial x_i}(x_i - x_i^0)$$ | Characterization of convexity for $C^1$-functions. |
| 12.18 | $f(\mathbf{x})$ is strictly convex on an open, convex subset $S$ in $R^n$ if and only if the inequality in (12.17) is strict for $\mathbf{x} \neq \mathbf{x}^0$. | Strict convexity for $C^1$-functions. |
| 12.19 | $$\mathbf{f}''(\mathbf{x}) = \begin{pmatrix} f''_{11}(\mathbf{x}) & f''_{12}(\mathbf{x}) & \cdots & f''_{1n}(\mathbf{x}) \\ f''_{21}(\mathbf{x}) & f''_{22}(\mathbf{x}) & \cdots & f''_{2n}(\mathbf{x}) \\ \vdots & \vdots & \ddots & \vdots \\ f''_{n1}(\mathbf{x}) & f''_{n2}(\mathbf{x}) & \cdots & f''_{nn}(\mathbf{x}) \end{pmatrix}$$ | The *Hessian matrix* of $\mathbf{f}$ at $\mathbf{x}$. |

12.20 The *principal minors* $\Delta_r(\mathbf{x})$ of order $r$ in the Hessian matrix $\mathbf{f}''(\mathbf{x})$ are the determinants of the sub-matrices obtained by deleting $n-r$ arbitrary rows and then deleting the $n-r$ columns having the same numbers.

Definition of the principal minors of the Hessian. (See also (18.21).)

12.21 $D_r(\mathbf{x}) = \begin{vmatrix} f''_{11}(\mathbf{x}) & f''_{12}(\mathbf{x}) & \cdots & f''_{1r}(\mathbf{x}) \\ f''_{21}(\mathbf{x}) & f''_{22}(\mathbf{x}) & \cdots & f''_{2r}(\mathbf{x}) \\ \vdots & \vdots & \ddots & \vdots \\ f''_{r1}(\mathbf{x}) & f''_{r2}(\mathbf{x}) & \cdots & f''_{rr}(\mathbf{x}) \end{vmatrix}$

The *leading principal minors* of the Hessian matrix of $f$ at $\mathbf{x}$, $r = 1, 2, \ldots, n$. (See also (18.22).)

12.22 $f(\mathbf{x})$ is concave on an open, convex subset $S$ in $R^n$ if and only if for all $\mathbf{x} \in S$ and for all $\Delta_r$,
$(-1)^r \Delta_r(\mathbf{x}) \geq 0$ for $r = 1, \ldots, n$.

Characterization of concavity for $C^2$-functions. ($\Delta_r$ defined in (12.20).)

12.23 $f(\mathbf{x})$ is strictly concave on an open, convex subset $S$ in $R^n$ if for all $\mathbf{x} \in S$,
$(-1)^r D_r(\mathbf{x}) > 0$ for all $r = 1, \ldots, n$.

Sufficient (but NOT necessary) conditions for strict concavity for $C^2$-functions. ($f(x) = -x^4$ is strictly concave, but $(-1)^1 D_1(0) = -f''(0) = 0$.)

12.24 $f(\mathbf{x})$ is convex on an open, convex subset $S$ in $R^n$ if and only if for all $\mathbf{x} \in S$ and all $\Delta_r$,
$\Delta_r(\mathbf{x}) \geq 0$ for $r = 1, \ldots, n$.

Characterization of convexity for $C^2$-functions. ($\Delta_r$ defined in (12.20).)

12.25 $f(\mathbf{x})$ is strictly convex on an open, convex subset $S$ in $R^n$ if for all $\mathbf{x} \in S$,
$D_r(\mathbf{x}) > 0$ for all $r = 1, \ldots, n$.

Sufficient (but NOT necessary) conditions for strict convexity for $C^2$-functions. ($f(x) = x^4$ is strictly convex, but $D_1(0) = f''(0) = 0$.)

12.26 $f(\mathbf{x})$ is *quasi-concave* on $S \subset R^n$ if the *upper level sets*
$P_a = \{\mathbf{x} \in S : f(\mathbf{x}) \geq a\}$
are convex for each real number $a$.

Definition of a quasi-concave function. (Upper level sets are also called *upper contour sets*.)

12.27 $f(\mathbf{x})$ is quasi-concave on an open, convex subset $S$ in $R^n$ if and only if
$f(\mathbf{x}) \geq f(\mathbf{x}^0) \Rightarrow f(\lambda \mathbf{x} + (1-\lambda)\mathbf{x}^0) \geq f(\mathbf{x}^0)$
for all $\mathbf{x}, \mathbf{x}^0 \in S$ and all $\lambda \in [0, 1]$.

Characterization of quasi-concavity.

| | | |
|---|---|---|
| 12.28 | $f(\mathbf{x})$ is *strictly quasi-concave* on an open, convex subset $S$ in $R^n$ if $$f(\mathbf{x}) \geq f(\mathbf{x}^0) \Rightarrow f(\lambda\mathbf{x} + (1-\lambda)\mathbf{x}^0) > f(\mathbf{x}^0)$$ for all $\mathbf{x} \neq \mathbf{x}^0 \in S$ and all $\lambda \in (0, 1)$. | The (most common) definition of strict quasi-concavity. |
| 12.29 | $f(\mathbf{x})$ is *(strictly) quasi-convex* on $S \subset R^n$ if $-f(\mathbf{x})$ is (strictly) quasi-concave. | Definition of a (strictly) quasi-convex function. |
| 12.30 | • $f(\mathbf{x})$ concave $\Rightarrow$ $f(\mathbf{x})$ quasi-concave. <br> • $f(\mathbf{x})$ convex $\Rightarrow$ $f(\mathbf{x})$ quasi-convex. <br> • A sum of quasi-concave functions is not necessarily quasi-concave. <br> • A sum of quasi-convex functions is not necessarily quasi-convex. <br> • If $f(\mathbf{x})$ is quasi-concave (quasi-convex) and $F$ is increasing, then $F(f(\mathbf{x}))$ is quasi-concave (quasi-convex). <br> • If $f(\mathbf{x})$ is quasi-concave (quasi-convex) and $F$ is decreasing, then $F(f(\mathbf{x}))$ is quasi-convex (quasi-concave). | Basic facts about quasi-convex and quasi-concave functions. |
| 12.31 | If $f_1, \ldots, f_m$ are concave functions defined on a convex set $S$ in $R^n$, and $g$ is defined over $S$ by $$g(\mathbf{x}) = F(f_1(\mathbf{x}), \ldots, f_m(\mathbf{x}))$$ with $F(u_1, \ldots, u_m)$ quasi-concave and increasing in each variable, then $g$ is quasi-concave. | A useful result. |
| 12.32 | $f(\mathbf{x})$ is quasi-concave on an open, convex set $S \subset R^n$ if and only if $$f(\mathbf{x}) \geq f(\mathbf{x}^0) \Rightarrow \sum_{i=1}^{n} \frac{\partial f(\mathbf{x}^0)}{\partial x_i}(x_i - x_i^0) \geq 0$$ for all $\mathbf{x}, \mathbf{x}^0 \in S$. | Characterization of quasi-concavity for $C^1$-functions. |
| 12.33 | Suppose all the first-order partials of $f(\mathbf{x})$ are different from zero in an open, convex set $S$ in $R^n$. Then $f(\mathbf{x})$ is quasi-concave on $S$ if and only if $$\sum_{i=1}^{n}\sum_{j=1}^{n} f''_{ij}(\mathbf{x}) h_i h_j \leq 0 \quad \text{when} \quad \nabla f(\mathbf{x}) \cdot \mathbf{h} = 0$$ for all $\mathbf{x} \in S$ and for all $\mathbf{h} = (h_1, \ldots, h_n)$. | Characterization of quasi-concavity for $C^2$-functions. |

12.34 $$B_r(\mathbf{x}) = \begin{vmatrix} 0 & f_1'(\mathbf{x}) & \cdots & f_r'(\mathbf{x}) \\ f_1'(\mathbf{x}) & f_{11}''(\mathbf{x}) & \cdots & f_{1r}''(\mathbf{x}) \\ \vdots & \vdots & \ddots & \vdots \\ f_r'(\mathbf{x}) & f_{r1}''(\mathbf{x}) & \cdots & f_{rr}''(\mathbf{x}) \end{vmatrix}$$ A *bordered Hessian* associated with $f$ at $\mathbf{x}$.

12.35 If $f(\mathbf{x})$ is quasi-concave on an open, convex subset $S$ of $R^n$, then for all $\mathbf{x} \in S$,
$(-1)^r B_r(\mathbf{x}) \geq 0$ for $r = 1, \ldots, n$.
A necessary condition for quasi-concavity of a $C^2$-function.

12.36 If $(-1)^r B_r(\mathbf{x}) > 0$ for $r = 1, \ldots, n$ for all $\mathbf{x}$ in an open, convex subset $S$ of $R^n$, then $f(\mathbf{x})$ is quasi-concave.
A sufficient condition for quasi-concavity of a $C^2$-function.

## References

See e.g. Nikaido (1968) and Takayama (1985).

# Chapter 13

# Classical optimization

| | | |
|---|---|---|
| 13.1 | $f(\mathbf{x}) = f(x_1, \ldots, x_n)$ has a *maximum (minimum)* at $\mathbf{x}^0 = (x_1^0, \ldots, x_n^0) \in S$ if $$f(\mathbf{x}^0) - f(\mathbf{x}) \geq 0 \ (\leq 0) \quad \text{for all } \mathbf{x} \in S$$ | Definition of (global) maximum (minimum) of a function of $n$ variables. A number $M$ which is either a maximum or a minimum of a function is called an *extremum*. |
| 13.2 | Suppose $f(\mathbf{x})$ is defined on $S \subset R^n$ and that $F(u)$ is strictly increasing on the range of $f$. Then $\mathbf{x}^0$ maximizes (minimizes) $f(\mathbf{x})$ on $S$ if and only if $\mathbf{x}^0$ maximizes (minimizes) $F(f(\mathbf{x}))$ on $S$. | An important fact. |
| 13.3 | If $f : S \to R$ is continuous on a closed, bounded subset $S$ of $R^n$, then there exist maximum and minimum points for $f$ in $S$, i.e. points $\mathbf{c}, \mathbf{d} \in S$ such that $$f(\mathbf{c}) \geq f(\mathbf{x}) \geq f(\mathbf{d}) \quad \text{for all } \mathbf{x} \in S$$ | The *extreme value theorem* (or *Weierstrass's theorem*). |
| 13.4 | $\mathbf{x}^0 = (x_1^0, \ldots, x_n^0)$ is a *stationary* point of $f(\mathbf{x})$ if $$f_1'(\mathbf{x}^0) = 0, \ f_2'(\mathbf{x}^0) = 0, \ldots, f_n'(\mathbf{x}^0) = 0$$ | Definition of stationary points for a function of $n$ variables. |
| 13.5 | If $f(\mathbf{x})$ has a maximum or minimum in $S \subset R^n$, then the maximum/minimum points are found among the following points:<br>• Interior points of $S$ that are stationary.<br>• Points at which $f$ is not differentiable.<br>• Extrema of $f$ at the boundary of $S$. | Where to find (global) maximum or minimum points. |
| 13.6 | Let $f(\mathbf{x})$ be concave and defined on a convex set $S$ in $R^n$, and let $\mathbf{x}^0$ be an interior point of $S$. Then $\mathbf{x}^0$ maximizes $f(\mathbf{x})$ if and only if $$f_1'(\mathbf{x}^0) = 0, \ f_2'(\mathbf{x}^0) = 0, \ldots, f_n'(\mathbf{x}^0) = 0$$ | Maximum of a concave function. |

| | | |
|---|---|---|
| 13.7 | Let $f(\mathbf{x})$ be convex and defined on a convex set $S$ in $R^n$, and let $\mathbf{x}^0$ be an interior point of $S$. Then $\mathbf{x}^0$ minimizes $f(\mathbf{x})$ if and only if $$f'_1(\mathbf{x}^0) = 0,\ f'_2(\mathbf{x}^0) = 0, \ldots, f'_n(\mathbf{x}^0) = 0$$ | Minimum of a convex function. |
| 13.8 | $f(\mathbf{x}) = f(x_1, \ldots, x_n)$ has a *local* maximum (minimum) at $\mathbf{x}^0$ if the inequality $\geq$ ($\leq$) in (13.1) is satisfied for every $\mathbf{x}$ in some $n$-ball $B(\mathbf{x}^0; r)$ contained in $S$. | Definition of local (or *relative*) maximum (minimum) points of a function of $n$ variables. A collective name is *local extrema*. |
| 13.9 | A stationary point $\mathbf{x}^0$ of $f(\mathbf{x}) = f(x_1, \ldots, x_n)$ is called a *saddle point* if it is neither a local maximum point nor a local minimum point, i.e. if every $n$-ball $B(\mathbf{x}^0; r)$ contains points $\mathbf{x}$ such that $f(\mathbf{x}) < f(\mathbf{x}^0)$ and other points $\mathbf{z}$ such that $f(\mathbf{z}) > f(\mathbf{x}^0)$. | Definition of a saddle point. |
| 13.10 | If $f(\mathbf{x}) = f(x_1, \ldots, x_n)$ has a local extremum at an interior point $\mathbf{x}^0$ of $S$, then $\mathbf{x}^0$ is a stationary point of $f$. | *First order conditions.* |
| 13.11 | If $f(\mathbf{x}) = f(x_1, \ldots, x_n)$ has a local maximum (minimum) at $\mathbf{x}^0$, then $$\sum_{i=1}^{n}\sum_{j=1}^{n} f''_{ij}(\mathbf{x}^0) h_i h_j \leq 0\ (\geq 0)$$ for all choices of $h_1, \ldots, h_n$. | A necessary (second-order) condition for local maximum (minimum). |
| 13.12 | $f'(x^0) = 0$ and $f''(x^0) < 0 \Longrightarrow$ $x^0$ is a local maximum point of $f$. $f'(x^0) = 0$ and $f''(x^0) > 0 \Longrightarrow$ $x^0$ is a local minimum point of $f$. | Sufficient conditions of local maximum (minimum) point for a function of one variable. |
| 13.13 | If $(x_0, y_0)$ is a stationary point of $f(x,y)$ and $D = f''_{11}(x_0,y_0) f''_{22}(x_0,y_0) - (f''_{12}(x_0,y_0))^2$, then <br> • $f''_{11}(x_0, y_0) > 0$ and $D > 0 \Longrightarrow$ $(x_0, y_0)$ is a local minimum point, <br> • $f''_{11}(x_0, y_0) < 0$ and $D > 0 \Longrightarrow$ $(x_0, y_0)$ is a local maximum point, <br> • $D < 0 \Longrightarrow (x_0, y_0)$ is a saddle point. | Classification of stationary points of a $C^2$-function. (Two variables). *Second-order conditions.* |

**13.14** If $\mathbf{x}^0 = (x_1^0, \ldots, x_n^0)$ is a stationary point of $f(x_1, \ldots, x_n)$, and if $D_k(\mathbf{x}^0)$ is the following determinant evaluated at $\mathbf{x}^0$,

$$D_k = \begin{vmatrix} f_{11}'' & \cdots & f_{1k}'' \\ \vdots & \ddots & \vdots \\ f_{k1}'' & \cdots & f_{kk}'' \end{vmatrix}, \quad k = 1, 2, \ldots, n$$

then
- if $(-1)^k D_k(\mathbf{x}^0) > 0$, $k = 1, \ldots, n$, then $\mathbf{x}^0$ is a local maximum point;
- if $D_k(\mathbf{x}^0) > 0$, $k = 1, \ldots, n$, then $\mathbf{x}^0$ is a local minimum point;
- if $D_n(\mathbf{x}^0) \neq 0$ and neither of the two conditions above is satisfied, then $\mathbf{x}^0$ is a saddle point.

*Classification of stationary points of a $C^2$-function of $n$ variables. Second-order conditions.*

**13.15** $\max f(\mathbf{x})$ when $\begin{cases} g_1(\mathbf{x}) = b_1 \\ \ldots\ldots\ldots \\ g_m(\mathbf{x}) = b_m \end{cases}$

*The Lagrange problem. $\mathbf{x} = (x_1, \ldots, x_n)$, and $m < n$.*

**13.16** *Lagrange's method* for solving problem (13.15): Introduce the Lagrangean

$$L(\mathbf{x}) = f(\mathbf{x}) - \sum_{j=1}^{m} \lambda_j (g_j(\mathbf{x}) - b_j)$$

where $\lambda_1, \ldots, \lambda_m$ are constants. Equate the partials of $L$ to 0, $k = 1, \ldots, n$:

$$\frac{\partial L(\mathbf{x})}{\partial x_k} = \frac{\partial f(\mathbf{x})}{\partial x_k} - \sum_{j=1}^{m} \lambda_j \frac{\partial g_j(\mathbf{x})}{\partial x_k} = 0$$

Solve these equations together with the $m$ constraints for $x_1, \ldots, x_n$ and $\lambda_1, \ldots, \lambda_m$.

*Necessary conditions for the solution of (13.15), with $f$ and $g_1, \ldots, g_m$ as $C^1$-functions. Assume the rank of the Jacobian $(\partial g_j / \partial x_i)_{m \times n}$ to be equal to $m$. (See (5.8).) The same conditions are necessary for the problem of minimizing $f(\mathbf{x})$ in (13.15).*

**13.17** If $\mathbf{x}^0$ is a solution to problem (13.15) and the gradients $\nabla g_1(\mathbf{x}^0), \ldots, \nabla g_m(\mathbf{x}^0)$ are linearly independent, then there exist unique numbers $\lambda_1, \ldots, \lambda_m$ such that

$$\nabla f(\mathbf{x}^0) = \lambda_1 \nabla g_1(\mathbf{x}^0) + \cdots + \lambda_m \nabla g_m(\mathbf{x}^0)$$

*An alternative formulation of (13.16).*

**13.18** Suppose $f(\mathbf{x})$ and $g_1(\mathbf{x}), \ldots, g_m(\mathbf{x})$ in (13.15) are defined on an open, convex set $S$ in $R^n$. Let $\mathbf{x}^0 \in S$ be a stationary point of the Lagrangean and suppose $g_j(\mathbf{x}^0) = b_j$, $j = 1, \ldots, m$. Then:

$L(\mathbf{x})$ concave $\Rightarrow \mathbf{x}^0$ solves problem (13.15).

*Sufficient conditions for the solution of problem (13.15). (For the minimization problem, replace $L(\mathbf{x})$ concave by $L(\mathbf{x})$ convex.)*

13.19 $$B_r = \begin{vmatrix} 0 & \cdots & 0 & \dfrac{\partial g_1}{\partial x_1} & \cdots & \dfrac{\partial g_1}{\partial x_r} \\ \vdots & \ddots & \vdots & \vdots & \ddots & \vdots \\ 0 & \cdots & 0 & \dfrac{\partial g_m}{\partial x_1} & \cdots & \dfrac{\partial g_m}{\partial x_r} \\ \dfrac{\partial g_1}{\partial x_1} & \cdots & \dfrac{\partial g_m}{\partial x_1} & F''_{11} & \cdots & F''_{1r} \\ \vdots & \ddots & \vdots & \vdots & \ddots & \vdots \\ \dfrac{\partial g_1}{\partial x_r} & \cdots & \dfrac{\partial g_m}{\partial x_r} & F''_{r1} & \cdots & F''_{rr} \end{vmatrix}$$

A *bordered Hessian* associated with (13.15), $r = 1, \ldots, n$.

Let $f(\mathbf{x})$ and $g_1(\mathbf{x}), \ldots, g_m(\mathbf{x})$ be $C^2$-functions in an open set $S$ in $R^n$ and let $\mathbf{x}^0 \in S$ satisfy the necessary conditions for problem (13.15) given in (13.16). Let $B_r(\mathbf{x}^0)$ be the determinant in (13.19) evaluated at $\mathbf{x}^0$. Then

13.20
- if $(-1)^m B_r(\mathbf{x}^0) > 0$, $r = m+1, \ldots, n$, then $\mathbf{x}^0$ is a local minimum point for problem (13.15);
- if $(-1)^r B_r(\mathbf{x}^0) > 0$, $r = m+1, \ldots, n$, then $\mathbf{x}^0$ is a local maximum point for problem (13.15).

*Local sufficient conditions* for the Lagrangean problem.

Suppose that $\mathbf{x}^0 = (x_1^0, \ldots, x_n^0)$ satisfies the necessary conditions in (13.16) for the problem

loc. max (min) $f(\mathbf{x})$ when $g(\mathbf{x}) = b$

Define

13.21 $$B_r = \begin{vmatrix} 0 & \dfrac{\partial g}{\partial x_1} & \cdots & \dfrac{\partial g}{\partial x_r} \\ \dfrac{\partial g}{\partial x_1} & F''_{11} & \cdots & F''_{1r} \\ \vdots & \vdots & \ddots & \vdots \\ \dfrac{\partial g}{\partial x_r} & F''_{r1} & \cdots & F''_{rr} \end{vmatrix}$$

Local second-order conditions for the case with one constraint.

Let $B_r(\mathbf{x}^0)$ be $B_r$ evaluated at $\mathbf{x}^0$. Then
- $\mathbf{x}^0$ is a local minimum point of $f$ subject to the given constraint if
  $B_r(\mathbf{x}^0) < 0$ for $r = 2, \ldots, n$
- $\mathbf{x}^0$ is a local maximum point of $f$ subject to the given constraint if
  $(-1)^r B_r(\mathbf{x}^0) > 0$ for $r = 2, \ldots, n$

| | | |
|---|---|---|
| 13.22 | $f^*(\mathbf{b}) = \max_{\mathbf{x}} \{f(\mathbf{x}) : g_j(\mathbf{x}) = b_j,\ j = 1,\ldots,m\}$ | The *value function* of problem (13.15), with $\mathbf{b} = (b_1,\ldots,b_m)$. |
| 13.23 | $\dfrac{\partial f^*(\mathbf{b})}{\partial b_i} = \lambda_i(\mathbf{b}), \qquad i = 1,\ldots,m$ | The Lagrangean multipliers as "shadow prices". $\lambda_1(\mathbf{b}),\ldots,\lambda_m(\mathbf{b})$ are the unique Lagrangian multipliers from (13.16) and (13.17). For more precise results, see Chapter 24. |

## References

Intriligator (1971), Luenberger (1973), and Dixit (1990).

# Chapter 14

# Linear and nonlinear programming

## Linear programming

14.1 $\displaystyle\max_{\mathbf{x}} \sum_{j=1}^{n} c_j x_j$ with $\begin{cases} \sum_{j=1}^{n} a_{1j}x_j \leq b_1 \\ \phantom{\sum_{j=1}^{n}} \cdots\cdots\cdots \\ \sum_{j=1}^{n} a_{mj}x_j \leq b_m \\ x_1 \geq 0, \ldots, x_n \geq 0 \end{cases}$

A *linear programming problem*. (The *primal problem*.) $\sum_{j=1}^{n} c_j x_j$ is called the *criterion function*. $(x_1, \ldots, x_n)$ is *admissible* if it satisfies all the $m+n$ constraints.

14.2 $\displaystyle\min_{\lambda} \sum_{i=1}^{m} b_i \lambda_i$ with $\begin{cases} \sum_{i=1}^{m} a_{i1}\lambda_i \geq c_1 \\ \phantom{\sum_{i=1}^{m}} \cdots\cdots\cdots \\ \sum_{i=1}^{m} a_{in}\lambda_i \geq c_n \\ \lambda_1 \geq 0, \ldots, \lambda_m \geq 0 \end{cases}$

The *dual* of (14.1). $\sum_{i=1}^{m} b_i \lambda_i$ is called the *criterion function*. $(\lambda_1, \ldots, \lambda_m)$ is *admissible* if it satisfies all the $n+m$ constraints.

14.3
$\max \mathbf{cx}$ with $\mathbf{Ax} \leq \mathbf{b}$, $\mathbf{x} \geq \mathbf{0}$
$\min \mathbf{b}'\lambda$ with $\mathbf{A}'\lambda \geq \mathbf{c}'$, $\lambda \geq \mathbf{0}$

A matrix formulation of (14.1) and (14.2).
$\mathbf{A} = (a_{ij})_{m \times n}$,
$\mathbf{x} = (x_j)_{n \times 1}$,
$\lambda = (\lambda_i)_{m \times 1}$,
$\mathbf{c} = (c_j)_{1 \times n}$,
$\mathbf{b} = (b_i)_{m \times 1}$.

14.4 If $(x_1, \ldots, x_n)$ is admissible in (14.1) and $(\lambda_1, \ldots, \lambda_m)$ is admissible in (14.2), then
$$b_1 \lambda_1 + \cdots + b_m \lambda_m \geq c_1 x_1 + \cdots + c_n x_n$$

The value of the criterion function in the dual is always greater than or equal to the value of the criterion function in the primal.

| | | |
|---|---|---|
| 14.5 | Suppose that $(x_1^*, \ldots, x_n^*)$ and $(\lambda_1^*, \ldots, \lambda_m^*)$ are admissible in (14.1) and (14.2) respectively, and that $$c_1 x_1^* + \cdots + c_n x_n^* = b_1 \lambda_1^* + \cdots + b_m \lambda_m^*$$ Then $(x_1^*, \ldots, x_n^*)$ and $(\lambda_1^*, \ldots, \lambda_m^*)$ are optimal in the respective problems. | An interesting result. |
| 14.6 | Suppose that the primal problem (14.1) has a (finite) optimal solution. Then the dual problem (14.2) also has a (finite) optimal solution and the values of the criterion functions are equal. If the primal has "unbounded maximum", then the dual has no admissible solutions. | The *duality theorem* in linear programming. |
| 14.7 | Consider problem (14.1). If we change $b_i$ to $b_i + \triangle b_i$ for $i = 1, \ldots, m$, and if the associated dual problems have the same optimal solution, $(\lambda_1^*, \ldots, \lambda_m^*)$, then the change in the optimal value of the criterion function of the primal problem is $$\triangle z^* = \lambda_1^* \triangle b_1 + \cdots + \lambda_m^* \triangle b_m$$ | An important sensitivity result. (The dual problems usually *have* the same solution if $\triangle b_1, \ldots, \triangle b_m$ are sufficiently small.) |
| 14.8 | The $i$th optimal dual variable $\lambda_i^*$ is equal to the change in criterion function of the primal problem (14.1) when $b_i$ is increased by one unit. | Interpretation of $\lambda_i^*$ as a *"shadow price"*. (A special case of (14.7), with the same qualifications.) |
| 14.9 | Suppose that the primal problem (14.1) has an optimal solution $(x_1^*, \ldots, x_n^*)$ and that the dual (14.2) has an optimal solution $(\lambda_1^*, \ldots, \lambda_m^*)$. Then for $i = 1, \ldots, n$, $j = 1, \ldots, m$: $$x_j^* > 0 \Rightarrow a_{1j}\lambda_1^* + \cdots + a_{mj}\lambda_m^* = c_j$$ $$\lambda_i^* > 0 \Rightarrow a_{i1}x_1^* + \cdots + a_{in}x_n^* = b_i$$ | *Complementary slackness.* |
| 14.10 | If the $i$th restriction in the primal problem (14.1) is an equality, then the $i$th dual variable has no sign restriction. | The case with equality restrictions in the primal problem. |
| 14.11 | If one of the variables in the primal problem (14.1) is without sign restriction, then the corresponding constraint in the dual is an equality. | The "dual statement" of (14.10). |

## Nonlinear programming

14.12 $\max_{\mathbf{x}} f(\mathbf{x})$ with $\begin{cases} g_1(\mathbf{x}) \leq b_1 \\ \ldots \ldots \ldots \\ g_m(\mathbf{x}) \leq b_m \end{cases}$

A *nonlinear programming problem*. $\mathbf{x} = (x_1, \ldots, x_n)$ is *admissible* if it satisfies all the constraints.

14.13 $\mathcal{L}(\mathbf{x}, \lambda) = f(\mathbf{x}) - \sum_{j=1}^{m} \lambda_j (g_j(\mathbf{x}) - b_j)$

The *Lagrangean function* associated with (14.12). $\lambda = (\lambda_1, \ldots, \lambda_m)$ are *Lagrangean multipliers* associated with the constraints.

14.14 Consider problem (14.12), where $f$ and $g_1, \ldots, g_m$ are $C^1$-functions. Suppose that there exist a vector $\lambda^0 = (\lambda_1^0, \ldots, \lambda_m^0)$ and a vector $\mathbf{x}^0 = (x_1^0, \ldots, x_n^0)$ such that

(a) $\dfrac{\partial \mathcal{L}(\mathbf{x}^0, \lambda^0)}{\partial x_i} = 0, \quad i = 1, \ldots, n$

(b) For all $j = 1, \ldots, m$, $\lambda_j^0 \geq 0$, and
$$\dfrac{\partial \mathcal{L}(\mathbf{x}^0, \lambda^0)}{\partial \lambda_j^0} \geq 0, \quad \lambda_j^0 \dfrac{\partial \mathcal{L}(\mathbf{x}^0, \lambda^0)}{\partial \lambda_j^0} = 0$$

(c) The Lagrangean function $\mathcal{L}(\mathbf{x}, \lambda)$ is a concave function of $\mathbf{x}$.

Then $\mathbf{x}^0$ solves problem (14.12).

The Kuhn-Tucker sufficient conditions. (The differentiable case.)

14.15 (b)' $\lambda_j^0 \geq 0$ $(= 0$ if $g_j(\mathbf{x}^0) < b_j)$, $j = 1, \ldots, m$

Alternative formulation of (b) in (14.14), assuming that $\mathbf{x}^0$ is admissible.

14.16 (14.14) is also valid if we replace (c) by the requirement

(c)' $f(\mathbf{x})$ is concave and $\lambda_j^0 g_j(\mathbf{x})$ is quasi-convex for $j = 1, \ldots, m$.

Alternative Kuhn-Tucker sufficient conditions.

14.17 Constraint $j$ in (14.12) is called *active at* $\mathbf{x}^0$ if $g_j(\mathbf{x}^0) = b_j$.

Definition of an *active* (or *binding*) constraint.

14.18 The following condition is often imposed in problem (14.12): The gradients at $\mathbf{x}^0$ of those $g_j$-functions whose constraints are active at $\mathbf{x}^0$ are linearly independent.

A *constraint qualification* for problem (14.12).

**14.19** Suppose that $\mathbf{x}^0 = (x_1^0, \ldots, x_n^0)$ solves (14.12) where $f$ and $g_1, \ldots, g_m$ are $C^1$-functions. Suppose further that the constraint qualification (14.18) is satisfied at $\mathbf{x}^0$. Then there exist unique numbers $\lambda_1, \ldots, \lambda_m$ such that

(a) $\dfrac{\partial \mathcal{L}(\mathbf{x}^0, \lambda)}{\partial x_i} = 0, \quad i = 1, \ldots, n$

(b) For all $j = 1, \ldots, m$, $\lambda_j \geq 0$, and
$$\dfrac{\partial \mathcal{L}(\mathbf{x}^0, \lambda)}{\partial \lambda_j} \geq 0, \quad \lambda_j \dfrac{\partial \mathcal{L}(\mathbf{x}^0, \lambda)}{\partial \lambda_j} = 0$$

The Kuhn-Tucker necessary conditions for problem (14.12). (Note that all the admissible points at which the constraint qualification fails to hold are candidates for optimality.)

**14.20** $(\mathbf{x}^0, \lambda^0)$ is a *saddle point* of the Lagrangean $\mathcal{L}(\mathbf{x}, \lambda)$ if for all $\lambda^0 \geq \mathbf{0}$ and all $\mathbf{x}$,
$$\mathcal{L}(\mathbf{x}, \lambda^0) \leq \mathcal{L}(\mathbf{x}^0, \lambda^0) \leq \mathcal{L}(\mathbf{x}^0, \lambda)$$

Definition of a saddle point for problem (14.12).

**14.21** If $\mathcal{L}(\mathbf{x}, \lambda)$ has a saddle point $(\mathbf{x}^0, \lambda^0)$, then $(\mathbf{x}^0, \lambda^0)$ solves problem (14.12).

Sufficient conditions for problem (14.12). (Note that no differentiability or concavity conditions are required.)

**14.22** The following condition is often imposed in problem (14.12): For some vector $\mathbf{x}' = (x_1', \ldots, x_n')$, $g_j(\mathbf{x}') < b_j$ for $j = 1, \ldots, m$.

The *Slater condition (qualification)*.

**14.23** Consider problem (14.12), assuming $f$ is concave and $g_1, \ldots, g_m$ are convex. Assume that the Slater condition (14.22) is satisfied. Then a necessary and sufficient condition for $\mathbf{x}^0$ to solve the problem is that there exist nonnegative numbers $\lambda_1^0, \ldots, \lambda_m^0$ such that $(\mathbf{x}^0, \lambda^0)$ is a saddle point of the Lagrangean $\mathcal{L}(\mathbf{x}, \lambda)$.

Saddle point result for concave programming. (It suffices that the conditions are satisfied in an open subset $X$ of $R^n$.)

**14.24** Consider problem (14.12) where $f$ and $g_1, \ldots, g_m$ are $C^1$-functions. Suppose that there exist numbers $\lambda_1, \ldots, \lambda_m$ and a vector $\mathbf{x}^0$ such that

- $\mathbf{x}^0$ satisfies (a) and (b) in (14.14).
- $\nabla f(\mathbf{x}^0) \neq \mathbf{0}$.
- $f(\mathbf{x})$ is quasi-concave and $\lambda_j g_j(\mathbf{x})$ is quasi-convex for $j = 1, \ldots, m$.

Then $\mathbf{x}^0$ solves problem (14.12).

Sufficient conditions for *quasi-concave programming*.

14.25 $\quad f^*(\mathbf{b}) = \max\{f(\mathbf{x}) : g_j(\mathbf{x}) \leq b_j,\ j = 1,\ldots,m\}$

The *value function* for (14.12), assuming that the maximum value exists.

14.26
- $f^*(\mathbf{b})$ is nondecreasing in each variable.
- $\partial f^*(\mathbf{b})/\partial b_j = \lambda_j(\mathbf{b}),\quad j = 1,\ldots,m.$
- If $f(\mathbf{x})$ is concave and $g_1(\mathbf{x}),\ \ldots,\ g_m(\mathbf{x})$ are convex, then $f^*(\mathbf{b})$ is concave.

Properties of the value function. (The second property is made precise in (24.14).)

## Nonlinear programming with nonnegativity conditions

14.27 $\quad \max_{\mathbf{x}} f(\mathbf{x})$ with $\begin{cases} g_1(\mathbf{x}) \leq b_1 \\ \cdots\cdots\cdots \\ g_m(\mathbf{x}) \leq b_m \end{cases},\quad \mathbf{x} \geq \mathbf{0}$

A nonlinear programming problem with explicit nonnegativity constraints on the variables. By writing the nonnegativity constraints as $g_{m+i}(\mathbf{x}) = -x_i \leq 0$ for $i = 1,\ldots,n$, we see that (14.27) is a special case of (14.12).

14.28 Consider problem (14.27), where $f$ and $g_1,\ldots,g_m$ are $C^1$-functions. Suppose that there exist numbers $\lambda_1^0,\ \ldots,\ \lambda_m^0$ and a vector $\mathbf{x}^0$ such that

(a) For all $i = 1,\ldots,n$, $x_i^0 \geq 0$ and
$$\frac{\partial \mathcal{L}(\mathbf{x}^0, \boldsymbol{\lambda}^0)}{\partial x_i} \leq 0,\quad x_i^0 \frac{\partial \mathcal{L}(\mathbf{x}^0, \boldsymbol{\lambda}^0)}{\partial x_i} = 0$$

(b) For all $j = 1,\ldots,m$, $\lambda_j^0 \geq 0$ and
$$\frac{\partial \mathcal{L}(\mathbf{x}^0, \boldsymbol{\lambda}^0)}{\partial \lambda_j^0} \geq 0,\quad \lambda_j^0 \frac{\partial \mathcal{L}(\mathbf{x}^0, \boldsymbol{\lambda}^0)}{\partial \lambda_j^0} = 0$$

(c) The Lagrangean function $\mathcal{L}(\mathbf{x}, \boldsymbol{\lambda})$ is a concave function of $\mathbf{x}$.

Then $\mathbf{x}^0$ solves problem (14.27).

Kuhn-Tucker sufficient conditions for problem (14.27).

14.29 (14.28) is also valid if we replace (c) by the requirement

(c)' $f(\mathbf{x})$ is concave and $\lambda_j^0 g_j(\mathbf{x})$ is quasi-convex for $j = 1,\ldots,m$.

Alternative Kuhn-Tucker sufficient conditions.

14.30 Suppose that $\mathbf{x}^0 = (x_1^0, \ldots, x_n^0)$ solves (14.27) where $f$ and $g_1, \ldots, g_m$ are $C^1$-functions. Suppose also that the gradients at $\mathbf{x}^0$ of those $g_j$ (including the functions $g_{m+1}, \ldots, g_{m+n}$ defined in the comment to (14.27)) whose constraints are active at $\mathbf{x}^0$ are linearly independent. Then there exist unique numbers $\lambda_1, \ldots, \lambda_m$ such that

(a) For all $i = 1, \ldots, n$, $x_i^0 \geq 0$, and
$$\frac{\partial \mathcal{L}(\mathbf{x}^0, \lambda^0)}{\partial x_i} \leq 0, \quad x_i^0 \frac{\partial \mathcal{L}(\mathbf{x}^0, \lambda^0)}{\partial x_i} = 0$$

(b) For all $j = 1, \ldots, m$, $\lambda_j^0 \geq 0$, and
$$\frac{\partial \mathcal{L}(\mathbf{x}^0, \lambda^0)}{\partial \lambda_j^0} \geq 0, \quad \lambda_j^0 \frac{\partial \mathcal{L}(\mathbf{x}^0, \lambda^0)}{\partial \lambda_j^0} = 0$$

*The Kuhn-Tucker necessary conditions for problem (14.27). (Note that all the admissible points at which the constraint qualification fails to hold are candidates for optimality.)*

14.31 Consider problem (14.27), assuming $f$ is concave and $g_1, \ldots, g_m$ are convex for all $\mathbf{x} \geq \mathbf{0}$. Assume that there exists a vector $\mathbf{x}' = (x_1', \ldots, x_n') \geq \mathbf{0}$ such that $g_j(\mathbf{x}') < b_j$ for $j = 1, \ldots, m$. Then a necessary and sufficient condition for $\mathbf{x}^0 \geq \mathbf{0}$ to solve the problem is that there exist numbers $\lambda_1^0, \ldots, \lambda_m^0$ such that

(a) $\mathcal{L}(\mathbf{x}, \lambda^0) \leq \mathcal{L}(\mathbf{x}^0, \lambda^0)$ for all $\mathbf{x} \geq \mathbf{0}$
(b) $\lambda_j^0 \geq 0$ (=0 if $g_j(\mathbf{x}^0) < b_j$), $j = 1, \ldots, m$
(c) $g_j(\mathbf{x}^0) \leq b_j$, $j = 1, \ldots, m$

*Kuhn-Tucker theorem for concave programming. (The non-differentiable case.)*

## References

Intriligator (1971), Luenberger (1973), Beavis and Dobbs (1990), and Dixit (1990).

# Chapter 15

# Calculus of variations and optimal control theory

## The calculus of variations

15.1 $\quad \max \int_{t_0}^{t_1} F(t, x, \dot{x}) \, dt, \quad x(t_0) = x_0, \quad x(t_1) = x_1$

The simplest problem in the calculus of variations. $F$ is a $C^2$-function of three variables. $x = x(t)$ is admissible if it is $C^1$ and satisfies the two boundary conditions. $t_0$, $t_1$, $x_0$, and $x_1$ are fixed numbers. To minimize the integral replace $F$ by $-F$.

15.2 $\quad \dfrac{\partial F}{\partial x} - \dfrac{d}{dt}\left(\dfrac{\partial F}{\partial \dot{x}}\right) = 0$

The *Euler equation*. A necessary condition for the solution of (15.1).

15.3 $\quad \dfrac{\partial^2 F}{\partial \dot{x} \partial \dot{x}} \cdot \ddot{x} + \dfrac{\partial^2 F}{\partial x \partial \dot{x}} \cdot \dot{x} + \dfrac{\partial^2 F}{\partial t \partial \dot{x}} - \dfrac{\partial F}{\partial x} = 0$

An alternative form of the Euler equation.

15.4 $\quad F''_{\dot{x}\dot{x}}(t, x(t), \dot{x}(t)) \leq 0 \quad \text{for all} \quad t \in [t_0, t_1]$

The *Legendre condition*. A necessary condition for the solution of (15.1).

15.5 If $F(t, x, \dot{x})$ is concave in $(x, \dot{x})$, an admissible function satisfying the Euler equation solves problem (15.1).

Sufficient conditions for the solution of (15.1).

15.6 $\quad x(t_1)$ free in (15.1) $\Rightarrow \left(\dfrac{\partial F}{\partial \dot{x}}\right)_{t=t_1} = 0$

*Transversality condition.* Adding condition (15.5) gives sufficient conditions.

| | | |
|---|---|---|
| 15.7 | $x(t_1) \geq x_1$ in (15.1) $\Rightarrow$ $$\left(\frac{\partial F}{\partial \dot{x}}\right)_{t=t_1} \leq 0 \,(= 0 \text{ if } x(t_1) > x_1)$$ | Transversality condition. Adding condition (15.5) gives sufficient conditions. |
| 15.8 | $t_1$ free in (15.1) $\Rightarrow$ $\left[F - \dot{x}\dfrac{\partial F}{\partial \dot{x}}\right]_{t=t_1} = 0$ | Transversality condition. |
| 15.9 | $x(t_1) = g(t_1)$ in (15.1) $\Rightarrow$ $$\left[F + (\dot{g} - \dot{x})\frac{\partial F}{\partial \dot{x}}\right]_{t=t_1} = 0$$ | Transversality condition. $g$ is a given $C^1$-function. |
| 15.10 | $\max\{\int_{t_0}^{t_1} F(t,x,\dot{x})\,dt + S(x(t_1))\}, \quad x(t_0) = x_0$ | A variational problem with a *scrap value function*, $S$, which is a $C^1$-function. |
| 15.11 | $\left(\dfrac{\partial F}{\partial \dot{x}}\right)_{t=t_1} + S'(x(t_1)) = 0$ | A solution to (15.10) must satisfy (15.2) and this transversality condition. |
| 15.12 | If $F(t,x,\dot{x})$ is concave in $(x,\dot{x})$ and $S(x)$ is concave, then an admissible function satisfying the Euler equation and (15.11) solves problem (15.10). | Sufficient conditions for the solution of (15.10). |
| 15.13 | $\max \int_{t_0}^{t_1} F\left(t, x, \dfrac{dx}{dt}, \dfrac{d^2x}{dt^2}, \ldots, \dfrac{d^n x}{dt^n}\right) dt$ | A variational problem with higher order derivatives. (Boundary conditions are unspecified.) |
| 15.14 | $\dfrac{\partial F}{\partial x} - \dfrac{d}{dt}\left(\dfrac{\partial F}{\partial \dot{x}}\right) + \ldots + (-1)^n \dfrac{d^n}{dt^n}\left(\dfrac{\partial F}{\partial x^{(n)}}\right) = 0$ | The *(generalized) Euler equation* for (15.13). |
| 15.15 | $\max \iint_R F\left(t, s, x, \dfrac{\partial x}{\partial t}, \dfrac{\partial x}{\partial s}\right) dt\,ds$ | A variational problem in which the unknown $x(t,s)$ is a function of two variables. (Boundary conditions are unspecified.) |
| 15.16 | $\dfrac{\partial F}{\partial x} - \dfrac{\partial}{\partial t}\left(\dfrac{\partial F}{\partial x'_t}\right) - \dfrac{\partial}{\partial s}\left(\dfrac{\partial F}{\partial x'_s}\right) = 0$ | The (generalized) Euler equation for (15.15). |

# Optimal control theory (continuous time)

15.17
$$\max_{\mathbf{u}} \int_{t_0}^{t_1} f_0(t, \mathbf{x}(t), \mathbf{u}(t))\, dt, \text{ when}$$
$$\dot{\mathbf{x}}(t) = \mathbf{f}(t, \mathbf{x}(t), \mathbf{u}(t)), \quad \mathbf{x}(t_0) = \mathbf{x}^0,$$
$$\mathbf{u}(t) = (u_1(t), \ldots, u_r(t)) \in U \subset R^r$$
(a) $x_i(t_1) = x_i^1$, $i = 1, \ldots, l$
(b) $x_i(t_1) \geq x_i^1$, $i = l+1, \ldots, q$
(c) $x_i(t_1)$ free, $i = q+1, \ldots, n$

A standard control problem with fixed time interval $[t_0, t_1]$. $\mathbf{f} = (f_1, \ldots, f_n)$. $\mathbf{x}(t) = (x_1(t), \ldots, x_n(t))$. $U$ is the control region, $\mathbf{u}(t)$ is piecewise continuous. To minimize the integral replace $f_0$ by $-f_0$.

15.18
$$H(t, \mathbf{x}, \mathbf{u}, \mathbf{p}) = \sum_{i=0}^{n} p_i f_i(t, \mathbf{x}, \mathbf{u})$$

The *Hamiltonian*.

15.19
If $(\mathbf{x}^*(t), \mathbf{u}^*(t))$ solves problem (15.17) then there exist a constant $p_0$ and a continuous and piecewise continuously differentiable function $\mathbf{p}(t) = (p_1(t), \ldots, p_n(t))$ such that for all $t \in [t_0, t_1]$:
- $p_0 = 0$ or $1$ and $(p_0, \mathbf{p}(t))$ is never $(0, \mathbf{0})$.
- For all $\mathbf{u} \in U$:
  $H(t, \mathbf{x}^*(t), \mathbf{u}, \mathbf{p}(t)) \leq H(t, \mathbf{x}^*(t), \mathbf{u}^*(t), \mathbf{p}(t))$
- $\dot{p}_i(t) = -\partial H^*/\partial x_i$, $i = 1, \ldots, n$,
- (a)' $p_i(t_1)$ no conditions, $i = 1, \ldots, l$
  (b)' $p_i(t_1) \geq 0$ ($= 0$ if $x_i^*(t_1) > x_i^1$),
  $\quad i = l+1, \ldots, q$
  (c)' $p_i(t_1) = 0$, $i = q+1, \ldots, n$

The *maximum principle*. $H^*$ denotes evaluation at $(t, \mathbf{x}^*(t), \mathbf{u}^*(t), \mathbf{p}(t))$. (The differential equation for $p_i(t)$ is not necessarily valid at the discontinuity points of $\mathbf{u}^*(t)$.)

15.20
If $(\mathbf{x}^*(t), \mathbf{u}^*(t))$ satisfies all the conditions in (15.19) for $p_0 = 1$, and if $H(t, \mathbf{x}, \mathbf{u}, \mathbf{p}(t))$ is concave in $(\mathbf{x}, \mathbf{u})$, then $(\mathbf{x}^*(t), \mathbf{u}^*(t))$ solves problem (15.17).

*Mangasarian's sufficiency conditions* for problem (15.17).

15.21 $\hat{H}(t, \mathbf{x}, \mathbf{p}) = \max\{H(t, \mathbf{x}, \mathbf{u}, \mathbf{p}) : \mathbf{u} \in U\}$

The *maximized* Hamiltonian.

15.22
The concavity condition in (15.20) can be replaced by the weaker condition that
$\hat{H}(t, \mathbf{x}, \mathbf{p}(t))$ is concave with respect to $\mathbf{x}$.

*Arrow's sufficiency condition.*

15.23 $\quad V(\mathbf{x}^0, \mathbf{x}^1, t_0, t_1) = \int_{t_0}^{t_1} f_0(t, \mathbf{x}^*(t), \mathbf{u}^*(t))\, dt$

The *value function* for problem (15.17), assuming that the solution is $(\mathbf{x}^*(t), \mathbf{u}^*(t))$ and that $\mathbf{x}^1 = (x_1^1, \ldots, x_q^1)$.

15.24
$$\frac{\partial V}{\partial x_i^0} = p_i(t_0), \quad i = 1, \ldots, n$$
$$\frac{\partial V}{\partial x_i^1} = -p_i(t_1), \quad i = 1, \ldots, q$$
$$\frac{\partial V}{\partial t_0} = -H^*(t_0), \quad \frac{\partial V}{\partial t_1} = H^*(t_1)$$

Properties of the value function. *Assume that $V$ is differentiable.* $H^*(t) = H(t, \mathbf{x}^*(t), \mathbf{u}^*(t), \mathbf{p}(t))$, (For precise assumptions, see Seierstad and Sydsæter (1987), section 3.5.)

15.25 If $t_1$ is free in problem (15.17) and $(\mathbf{x}^*(t), \mathbf{u}^*(t))$ solves the corresponding problem on $[t_0, t_1^*]$, then all the conditions in (15.19) are satisfied on $[t_0, t_1^*]$, and in addition

$$H(t_1^*, \mathbf{x}^*(t_1^*), \mathbf{u}^*(t_1^*), \mathbf{p}(t_1^*)) = 0$$

Necessary conditions for a free terminal time problem. (For sufficient conditions, see Seierstad and Sydsæter (1987), section 2.9.)

15.26
$$\max_{\mathbf{u}} \left[ \int_{t_0}^{t_1} f_0(t, \mathbf{x}, \mathbf{u}) e^{-rt}\, dt + S(t_1, \mathbf{x}(t_1)) e^{-rt_1} \right]$$
$$\dot{\mathbf{x}} = \mathbf{f}(t, \mathbf{x}, \mathbf{u}), \quad \mathbf{x}(t_0) = \mathbf{x}^0$$
$$\mathbf{u} = (u_1, \ldots, u_r) \in U$$

A control problem with a *scrap value function*. ($t_1$ fixed.)

15.27 $\quad H^c(t, \mathbf{x}, \mathbf{u}, \mathbf{q}) = \sum_{i=0}^{n} q_i f_i(t, \mathbf{x}, \mathbf{u})$

The *current value Hamiltonian*.

15.28 If $(\mathbf{x}^*(t), \mathbf{u}^*(t))$ solves problem (15.26), there exists a continuous and piecewise continuously differentiable function $\mathbf{q}(t) = (q_1(t), \ldots, q_n(t))$ such that for all $t \in [t_0, t_1]$, with $q_0 = 1$:
- For all $\mathbf{u} \in U$,
$$H^c(t, \mathbf{x}^*(t), \mathbf{u}, \mathbf{q}(t)) \leq H^c(t, \mathbf{x}^*(t), \mathbf{u}^*(t), \mathbf{q}(t))$$
- $\dot{q}_i - rq_i = -\dfrac{\partial H^c(t, \mathbf{x}^*, \mathbf{u}^*, \mathbf{q})}{\partial x_i}, \quad i = 1, \ldots, n$
- $q_i(t_1) = \dfrac{\partial S(t_1, \mathbf{x}^*(t_1))}{\partial x_i}, \quad i = 1, \ldots, n$

The maximum principle for problem (15.26); current value formulation. (The differential equation for $q_i(t)$ is not necessarily valid at the discontinuity points of $\mathbf{u}^*(t)$.)

| | | |
|---|---|---|
| 15.29 | If $(\mathbf{x}^*(t), \mathbf{u}^*(t))$ satisfies the conditions in (15.28), if $H^c(t, \mathbf{x}, \mathbf{u}, \mathbf{q}(t))$ is concave in $(\mathbf{x}, \mathbf{u})$, and if $S(t, \mathbf{x})$ is concave in $\mathbf{x}$, then $(\mathbf{x}^*(t), \mathbf{u}^*(t))$ solves the problem. | Sufficient conditions for the solution of (15.26). |
| 15.30 | $\hat{H}^c(t, \mathbf{x}, \mathbf{q}) = \max\{H^c(t, \mathbf{x}, \mathbf{u}, \mathbf{q}) : \mathbf{u} \in U\}$ | The *maximized current value Hamiltonian*. |
| 15.31 | The concavity condition on $H^c$ in (15.29) can be replaced by the condition that $\hat{H}^c(t, \mathbf{x}, \mathbf{q}(t))$ is concave in $\mathbf{x}$. | The *Arrow sufficiency condition*. |
| 15.32 | If $t_1$ is free in problem (15.26), and if $(\mathbf{x}^*, \mathbf{u}^*)$ solves the corresponding problem on $[t_0, t_1^*]$, then all the conditions in (15.28) are satisfied on $[t_0, t_1^*]$ and in addition $$H^c(t_1^*, \mathbf{x}^*(t_1^*), \mathbf{u}^*(t_1^*), \mathbf{q}(t_1^*)) = rS(t_1^*, \mathbf{x}^*(t_1^*)) - \frac{\partial S(t_1^*, \mathbf{x}^*(t_1^*))}{\partial t_1}$$ | Necessary conditions for problem (15.26) when $t_1$ is free. |
| 15.33 | $$\min_{\mathbf{u}} \left[ \int_{t_0}^{t_1} (\mathbf{x}'\mathbf{A}\mathbf{x} + \mathbf{u}'\mathbf{B}\mathbf{u})\, dt + (\mathbf{x}(t_1))'\mathbf{S}\mathbf{x}(t_1) \right]$$ $\dot{\mathbf{x}} = \mathbf{F}\mathbf{x} + \mathbf{G}\mathbf{u}, \quad \mathbf{x}(t_0) = \mathbf{x}^0, \quad \mathbf{u} \in R^r$ $\mathbf{A} = \mathbf{A}(t)_{n\times n}$ and $\mathbf{S}_{n\times n}$ are symmetric and positive semidefinite, $\mathbf{B} = \mathbf{B}(t)_{r\times r}$ is symmetric and positive definite. $\mathbf{F} = \mathbf{F}(t)_{n\times n}$ and $\mathbf{G} = \mathbf{G}(t)_{n\times r}$. | A *linear quadratic control problem*. The entries of $\mathbf{A}(t), \mathbf{B}(t), \mathbf{F}(t)$, and $\mathbf{G}(t)$ are continuous functions of $t$. $\mathbf{x} = \mathbf{x}(t)$ is $n \times 1$, $\mathbf{u} = \mathbf{u}(t)$ is $r \times 1$. |
| 15.34 | $\dot{\mathbf{R}} = -\mathbf{R}\mathbf{F} - \mathbf{F}'\mathbf{R} + \mathbf{R}\mathbf{G}\mathbf{B}^{-1}\mathbf{G}'\mathbf{R} - \mathbf{A}$ | The *Riccati* equation associated with (15.33). |
| 15.35 | If $(\mathbf{x}^*(t), \mathbf{u}^*(t))$ is an admissible pair in problem (15.33), and $\mathbf{u}^* = -(\mathbf{B}(t))^{-1}(\mathbf{G}(t))'\mathbf{R}(t)\mathbf{x}^*$, with $\mathbf{R} = \mathbf{R}(t)$ as a symmetric $n \times n$-matrix with $C^1$-entries satisfying (15.34) with $\mathbf{R}(t_1) = \mathbf{S}$, then $(\mathbf{x}^*(t), \mathbf{u}^*(t))$ solves problem (15.33). | The solution of (15.33). |

## Infinite horizon

15.36
$$\max \int_{t_0}^{\infty} f_0(t, \mathbf{x}(t), \mathbf{u}(t))\, dt$$
$$\dot{\mathbf{x}}(t) = \mathbf{f}(t, \mathbf{x}(t), \mathbf{u}(t)),\ \mathbf{x}(t_0) = \mathbf{x}^0,\ \mathbf{u}(t) \in U$$
$$\lim_{t \to \infty} \mathbf{x}(t) = \mathbf{x}^1, \quad (\mathbf{x}^1 \text{ fixed vector in } R^n)$$

The simplest *infinite horizon problem*, assuming that the integral converges for all admissible pairs.

15.37
$$D(t) = \int_{t_0}^{t} (f_0^* - f_0)\, d\tau \quad \text{where}$$
$$f_0^* = f_0(\tau, \mathbf{x}^*(\tau), \mathbf{u}^*(\tau)),\ f_0 = f_0(\tau, \mathbf{x}(\tau), \mathbf{u}(\tau))$$

$(\mathbf{x}^*(t), \mathbf{u}^*(t))$ is a candidate for optimality, and $(\mathbf{x}(t), \mathbf{u}(t))$ is any admissible pair.

15.38 $(\mathbf{x}^*(t), \mathbf{u}^*(t))$ is *sporadically catching up optimal* (SCU-optimal) if for every admissible pair $(\mathbf{x}(t), \mathbf{u}(t))$, $\overline{\lim}_{t \to \infty} D(t) \geq 0$, e.g. for every $\varepsilon > 0$ and every $T$ there is some $t \geq T$ such that $D(t) \geq -\varepsilon$.

Different optimality criteria for infinite horizon problems.

15.39 $(\mathbf{x}^*(t), \mathbf{u}^*(t))$ is *catching up optimal* (CU-optimal) if for every admissible pair $(\mathbf{x}(t), \mathbf{u}(t))$, $\underline{\lim}_{t \to \infty} D(t) \geq 0$, e.g. for every $\varepsilon > 0$ there is some $T$ such that if $t \geq T$, then $D(t) \geq -\varepsilon$.

15.40 $(\mathbf{x}^*(t), \mathbf{u}^*(t))$ is *overtaking optimal* (OT-optimal) if for every admissible pair $(\mathbf{x}(t), \mathbf{u}(t))$, there exist a number $T$ such that $D(t) \geq 0$ for all $t \geq T$.

15.41 OT-optimality $\Longrightarrow$ CU-optimality $\Longrightarrow$ SCU-optimality

Relationship between the optimality criteria.

15.42
$$\lim_{t \to \infty} x_i(t) = x_i^1,\ i = 1, \ldots, l$$
$$\lim_{t \to \infty} x_i(t) \geq x_i^1,\ i = l+1, \ldots, q$$

Terminal conditions.

15.43 A pair $(\mathbf{x}^*(t), \mathbf{u}^*(t))$ which satisfies the differential equation in (15.36) and the terminal conditions (15.42), and which is SCU-, CU-, or OT-optimal, must satisfy all the conditions in (15.19) on $[0, \infty)$, except for the transversality conditions.

The maximum principle. Infinite horizon.

15.44
In the case of CU-optimality, the conditions in (15.43) are sufficient for optimality if

- $H(t, \mathbf{x}, \mathbf{u}, \mathbf{p}(t))$ is concave in $(\mathbf{x}, \mathbf{u})$
- For all admissible $\mathbf{x}(t)$,
$$\lim_{t \to \infty} \mathbf{p}(t) \cdot (\mathbf{x}(t) - \mathbf{x}^*(t)) \geq 0$$

*Sufficiency conditions* for the infinite horizon case.

## References

Most of the definitions and results are (in a slightly less general form) in Kamien and Schwartz (1991). For more details, see e.g. Seierstad and Sydsæter (1987) or Feichtinger and Hartl (1986) (in German).

# Chapter 16

# Discrete dynamic optimization

## Dynamic programming

16.1
$$\max_{\mathbf{u}_t} \sum_{t=0}^{T} f_0(t, \mathbf{x}_t, \mathbf{u}_t)$$
$$\mathbf{x}_{t+1} = \mathbf{f}(t, \mathbf{x}_t, \mathbf{u}_t), \quad t = 0, \ldots, T-1$$
$$\mathbf{x}_0 = \mathbf{x}^0, \quad \mathbf{u}_t \in U$$

A *dynamic programming problem*. $\mathbf{x}_t$ is an $n$-vector, $\mathbf{u}_t$ is an $r$-vector, $\mathbf{f} = (f_1, \ldots, f_n)$, $U$ is a subset of $R^r$ and $\mathbf{x}^0$ is a fixed vector in $R^n$.

16.2
$$J_s(\mathbf{x}) = \max_{\mathbf{u}_s, \ldots, \mathbf{u}_T \in U} \sum_{t=s}^{T} f_0(t, \mathbf{x}_t, \mathbf{u}_t), \text{ where}$$
$$\mathbf{x}_{t+1} = \mathbf{f}(t, \mathbf{x}_t, \mathbf{u}_t), \quad t = s, \ldots, T-1, \mathbf{x}_s = \mathbf{x}$$

Definition of the *value function*, $J_s(\mathbf{x})$, for problem (16.1).

16.3
$$J_T(\mathbf{x}) = \max_{\mathbf{u} \in U} f_0(T, \mathbf{x}, \mathbf{u})$$
$$J_s(\mathbf{x}) = \max_{\mathbf{u} \in U} \left[ f_0(s, \mathbf{x}, \mathbf{u}) + J_{s+1}(\mathbf{f}(s, \mathbf{x}, \mathbf{u})) \right]$$
for $s = 0, 1, \ldots, T-1$

The *fundamental equations* in dynamic programming.

16.4
$$\max \sum_{t=0}^{\infty} g(\mathbf{x}_t, \mathbf{u}_t) \alpha^t, \quad \mathbf{u}_t \in U \subset R^r$$
$$\mathbf{x}_{t+1} = \mathbf{f}(\mathbf{x}_t, \mathbf{u}_t), \quad \mathbf{x}_0 = \mathbf{x}^0, \quad t = 0, 1, 2, \ldots$$

An infinite horizon problem. $g$ and $\mathbf{f}$ are functions of $n + r$ variables, $\alpha \in (0, 1)$ is a constant discount factor.

16.5 The sequence of pairs $\{(\mathbf{x}_t, \mathbf{u}_t)\}$ is called *admissible* provided $\mathbf{u}_t \in U$, $\mathbf{x}_0 = \mathbf{x}^0$ and the difference equation in (16.4) is satisfied for all $t = 0, 1, 2, \ldots$

Definition of an *admissible* sequence.

| | | |
|---|---|---|
| 16.6 | (B) $M_1 \leq g(\mathbf{x}, \mathbf{u}) \leq M_2$<br>(BB) $g(\mathbf{x}, \mathbf{u}) \geq M$<br>(BA) $g(\mathbf{x}, \mathbf{u}) \leq N$ | Boundedness conditions. $M_1$, $M_2$, and $M$ are given numbers. |
| 16.7 | $V(\mathbf{x}, \pi, s, \infty) = \sum_{t=s}^{\infty} g(\mathbf{x}_t, \mathbf{u}_t)\alpha^t$, where $\pi = (\mathbf{u}_s, \mathbf{u}_{s+1}, \ldots)$, with $\mathbf{u}_{s+k} \in U$ for $k = 0, 1, \ldots$, and $\mathbf{x}_{t+1} = f(\mathbf{x}_t, \mathbf{u}_t)$ for $t = s, s+1, \ldots$, $\mathbf{x}_s = \mathbf{x}$. | The *total utility* obtained from period $s$ and onwards. |
| 16.8 | $J_s(\mathbf{x}) = \sup_\pi V(\mathbf{x}, \pi, s, \infty)$ where the supremum is taken over all $\pi = (\mathbf{u}_s, \mathbf{u}_{s+1}, \ldots)$ with $\mathbf{u}_{s+k} \in U$, with $(\mathbf{x}_t, \mathbf{u}_t)$ admissible for $t \geq s$, and with $\mathbf{x}_s = \mathbf{x}$. | The *value function* for problem (16.4). |
| 16.9 | $J_s(\mathbf{x}) = \alpha^s J_0(\mathbf{x})$, $s = 1, 2, \ldots$<br>$J_0(\mathbf{x}) = \max_{\mathbf{u} \in U}\{g(\mathbf{x}, \mathbf{u}) + \alpha J_0(f(\mathbf{x}, \mathbf{u}))\}$ | Properties of the value function, assuming that one of the boundedness conditions in (16.6) is satisfied. |

## Discrete optimal control theory

| | | |
|---|---|---|
| 16.10 | $H = f_0(t, \mathbf{x}, \mathbf{u}) + \mathbf{p}f(t, \mathbf{x}, \mathbf{u})$, $t = 0, \ldots, T$ | The Hamiltonian $H = H(t, \mathbf{x}, \mathbf{u}, \mathbf{p})$ associated with (16.1), with $\mathbf{p} = (p^1, \ldots, p^n)$. |
| 16.11 | Suppose $\{(\mathbf{x}_t^*, \mathbf{u}_t^*)\}$ is an optimal sequence for problem (16.1). Then there exist vectors $\mathbf{p}_t \in R^n$ such that for $t = 0, \ldots, T$,<br>$H'_\mathbf{u}(t, \mathbf{x}_t^*, \mathbf{u}_t^*, \mathbf{p}_t) \cdot (\mathbf{u} - \mathbf{u}_t^*) \leq 0$ for all $\mathbf{u} \in U$<br>The vector $\mathbf{p}_t = (p_t^1, \ldots, p_t^n)$ is a solution of<br>$\mathbf{p}_{t-1} = H'_\mathbf{x}(t, \mathbf{x}_t^*, \mathbf{u}_t^*, \mathbf{p}_t)$, $t = 1, \ldots, T$<br>with $\mathbf{p}_T = \mathbf{0}$. | The maximum principle for (16.1). Necessary conditions for optimality. $U$ is convex. |
| 16.12 | (a) $x_T^i = \bar{x}^i$ for $i = 1, \ldots, l$<br>(b) $x_T^i \geq \bar{x}^i$ for $i = l+1, \ldots, m$<br>(c) $x_T^i$ free for $i = m+1, \ldots, n$ | Terminal conditions for problem (16.1). |
| 16.13 | $H = \begin{cases} q_0 f_0(t, \mathbf{x}, \mathbf{u}) + \mathbf{p}f(t, \mathbf{x}, \mathbf{u}), & t = 0, \ldots, T-1 \\ q_0 f_0(T, \mathbf{x}, \mathbf{u}), & t = T \end{cases}$ | The Hamiltonian $H = H(t, \mathbf{x}, \mathbf{u}, \mathbf{p})$ associated with (16.1) with terminal conditions (16.12). |

16.14  Suppose $\{(\mathbf{x}_t^*, \mathbf{u}_t^*)\}$ is an optimal sequence for problem (16.1) with terminal condition (16.12). Then there exist vectors $\mathbf{p}_t \in R^n$ and a number $q_0$, with $(q_0, \mathbf{p}_T) \neq (0, \mathbf{0})$ and with $q_0 = 0$ or $1$, such that for $t = 0, \ldots, T$,
$$H'_\mathbf{u}(t, \mathbf{x}_t^*, \mathbf{u}_t^*, \mathbf{p}_t)(\mathbf{u} - \mathbf{u}_t^*) \leq 0 \text{ for all } \mathbf{u} \in U$$
The vector $\mathbf{p}_t = (p_t^1, \ldots, p_t^n)$ is a solution of
$$p_{t-1}^i = H'_{x^i}(t, \mathbf{x}_t^*, \mathbf{u}_t^*, \mathbf{p}_t), \quad t = 1, \ldots, T-1$$
Moreover,
$$p_{T-1}^i = q_0 \frac{\partial f_0(T, \mathbf{x}_T^*, \mathbf{u}_T^*)}{\partial x^i} + p_T^i$$
where $p_T^i$ satisfies

(a)'  $p_T^i$ no condition $\quad i = 1, \ldots, l$
(b)'  $p_T^i \geq 0 \ (= 0 \text{ if } x_T^{*i} > \bar{x}^i) \quad i = l+1, \ldots, m$
(c)'  $p_T^i = 0 \quad i = m+1, \ldots, n$

The maximum principle for (16.1) with terminal conditions (16.12). Necessary conditions for optimality. $U$ is convex.

16.15  Suppose that the sequence $\{(\mathbf{x}_t^*, \mathbf{u}_t^*, \mathbf{p}_t)\}$ satisfies all the conditions in (16.14) for $q_0 = 1$, and suppose further that $H(t, \mathbf{x}, \mathbf{u}, \mathbf{p}_t)$ is concave with respect to $(\mathbf{x}, \mathbf{u})$ for every $t$. Then the sequence $\{(\mathbf{x}_t^*, \mathbf{u}_t^*, \mathbf{p}_t)\}$ is optimal.

Sufficient conditions for optimality.

## References

See Bellman (1957) and Varaiya (1972).

# Chapter 17

# Vectors. Linear dependence. Scalar products

| | | |
|---|---|---|
| 17.1 | $\mathbf{a}_1 = \begin{pmatrix} a_{11} \\ a_{21} \\ \vdots \\ a_{m1} \end{pmatrix}, \ldots, \mathbf{a}_n = \begin{pmatrix} a_{1n} \\ a_{2n} \\ \vdots \\ a_{mn} \end{pmatrix}$ | $n$ (column-) vectors in $R^m$. |
| 17.2 | If $x_1, x_2, \ldots, x_n$ are real numbers, then $x_1\mathbf{a}_1 + x_2\mathbf{a}_2 + \cdots + x_n\mathbf{a}_n$ is a *linear combination* of the vectors $\mathbf{a}_1, \mathbf{a}_2, \ldots, \mathbf{a}_n$ | Definition of a linear combination of vectors. |
| 17.3 | $\mathbf{a}_1, \mathbf{a}_2, \ldots, \mathbf{a}_n$ in $R^m$ are *linearly dependent* if there exist numbers $c_1, c_2, \ldots, c_n$, not all zero, such that $c_1\mathbf{a}_1 + c_2\mathbf{a}_2 + \cdots + c_n\mathbf{a}_n = \mathbf{0}$ | Definition of linear dependence. |
| 17.4 | $\mathbf{a}_1, \mathbf{a}_2, \ldots, \mathbf{a}_n$ in $R^m$ are *linearly independent* if they are not linearly dependent i.e. if $c_1\mathbf{a}_1 + c_2\mathbf{a}_2 + \cdots + c_n\mathbf{a}_n = \mathbf{0}$ implies that all $c_1, c_2, \ldots, c_n$ are zero. | Definition of linear independence. |
| 17.5 | The vectors $\mathbf{a}_1, \mathbf{a}_2, \ldots, \mathbf{a}_n$ in (17.1) are linearly independent if and only if the matrix $(a_{ij})_{m \times n}$ has rank $n$. | A characterization of linear independence for $n$ vectors in $R^m$. (See (19.22) for the definition of rank.) |
| 17.6 | The vectors $\mathbf{a}_1, \mathbf{a}_2, \ldots, \mathbf{a}_n$ in $R^n$ are linearly independent if and only if $\begin{vmatrix} a_{11} & a_{12} & \cdots & a_{1n} \\ a_{21} & a_{22} & \cdots & a_{2n} \\ \vdots & \vdots & \ddots & \vdots \\ a_{n1} & a_{n2} & \cdots & a_{nn} \end{vmatrix} \neq 0$ | A characterization of linear independence for $n$ vectors in $R^n$. (A special case of (17.5).) |

| | | |
|---|---|---|
| 17.7 | A non-empty subset $V$ of vectors in $R^m$ is a *subspace* of $R^m$ if $\mathbf{a}_1, \mathbf{a}_2 \in V$ and $c_1, c_2$ arbitrary numbers imply $c_1\mathbf{a}_1 + c_2\mathbf{a}_2 \in V$. | Definition of a subspace. |
| 17.8 | $\mathcal{S}[\mathbf{a}_1, \ldots, \mathbf{a}_n]$ is the subspace of vectors which can be written as linear combinations of $\mathbf{a}_1, \ldots, \mathbf{a}_n$. | Definition of the *span* of $n$ vectors $\mathbf{a}_1, \ldots, \mathbf{a}_n$ in $R^m$. |
| 17.9 | A collection of vectors $\mathbf{a}_1, \ldots, \mathbf{a}_n$ belonging to a subspace $V$ of $R^m$ is a *basis* for $V$ if<br>• $\mathbf{a}_1, \ldots, \mathbf{a}_n$ are linearly independent<br>• $\mathcal{S}[\mathbf{a}_1, \ldots, \mathbf{a}_n] = V$ | Definition of a basis for a subspace. |
| 17.10 | The *dimension* of a subspace $V$ of $R^m$ is the number of vectors in a basis for $V$. (Two bases for $V$ have always the same number of vectors.) | Definition of the dimension of a subspace. |
| 17.11 | Let a subspace $V$ of $R^m$ have dimension $n$.<br>• Any collection of $n$ linearly independent vectors in $V$ is a basis for $V$.<br>• Any collection of $n$ vectors in $V$ which spans $V$ is a basis for $V$. | Important facts about subspaces. |
| 17.12 | The *scalar product* of $\mathbf{a} = (a_1, \ldots, a_m)$ and $\mathbf{b} = (b_1, \ldots, b_m)$ is the number<br>$$\mathbf{a} \cdot \mathbf{b} = a_1 b_1 + \cdots + a_m b_m$$ | Definition of the scalar product, also called *inner product* or *dot product*. |
| 17.13 | $\mathbf{a} \cdot \mathbf{b} = \mathbf{b} \cdot \mathbf{a}$<br>$\mathbf{a} \cdot (\mathbf{b} + \mathbf{c}) = \mathbf{a} \cdot \mathbf{b} + \mathbf{a} \cdot \mathbf{c}$<br>$(\alpha \mathbf{a}) \cdot \mathbf{b} = \mathbf{a} \cdot (\alpha \mathbf{b}) = \alpha (\mathbf{a} \cdot \mathbf{b})$<br>$\mathbf{a} \cdot \mathbf{a} > 0 \iff \mathbf{a} \neq \mathbf{0}$ | Properties of the scalar product. $\alpha$ is a real (or complex) number. |
| 17.14 | $\|\mathbf{a}\| = \sqrt{a_1^2 + a_2^2 + \cdots + a_m^2}$ | Definition of the *norm* (or length) of a vector $\mathbf{a} = (a_1, a_2, \ldots, a_m)$. |
| 17.15 | (a) $\|\mathbf{a}\| > 0$ for $\mathbf{a} \neq \mathbf{0}$ and $\|\mathbf{0}\| = 0$<br>(b) $\|\alpha \mathbf{a}\| = \|\alpha\| \|\mathbf{a}\|$<br>(c) $\|\mathbf{a} + \mathbf{b}\| \leq \|\mathbf{a}\| + \|\mathbf{b}\|$<br>(d) $\|\mathbf{a} \cdot \mathbf{b}\| \leq \|\mathbf{a}\| \cdot \|\mathbf{b}\|$ | Properties of the norm. $\mathbf{a}$ and $\mathbf{b}$ are vectors in $R^m$, $\alpha$ is a real (or complex) number. (d) is called the *Schwartz's inequality*. |

| | | |
|---|---|---|
| 17.16 | $\|\mathbf{a}-\mathbf{b}\| = \sqrt{(a_1-b_1)^2+\cdots+(a_m-b_m)^2}$ | The *distance* between $\mathbf{a}=(a_1,\ldots,a_m)$ and $\mathbf{b}=(b_1,\ldots,b_m)$ in $R^m$. |
| 17.17 | The *angle* $\phi$ between two vectors $\mathbf{a}$ and $\mathbf{b}$ is defined by $$\cos\phi = \frac{\mathbf{a}\cdot\mathbf{b}}{\|\mathbf{a}\|\cdot\|\mathbf{b}\|}, \qquad 0°\le\phi\le 180°$$ | Definition of the angle between two vectors in $R^m$. (Because of (17.15) (d), $|\cos\phi|\le 1$.) |

## References

All the formulas are standard and are found in almost any linear algebra text, e.g. Anton (1987) or Lang (1987).

# Chapter 18

## Determinants

18.1 $\begin{vmatrix} a_{11} & a_{12} \\ a_{21} & a_{22} \end{vmatrix} = a_{11}a_{22} - a_{21}a_{12}$  Definition of a 2 × 2-determinant.

18.2 $\begin{vmatrix} a_{11} & a_{12} & a_{13} \\ a_{21} & a_{22} & a_{23} \\ a_{31} & a_{32} & a_{33} \end{vmatrix} = \begin{cases} a_{11}a_{22}a_{33} - a_{11}a_{23}a_{32} \\ + a_{12}a_{23}a_{31} - a_{12}a_{21}a_{33} \\ + a_{13}a_{21}a_{32} - a_{13}a_{22}a_{31} \end{cases}$  Definition of a 3 × 3-determinant.

18.3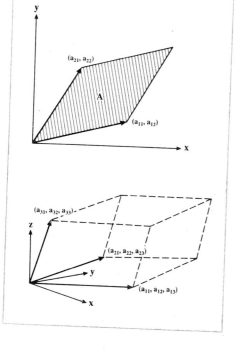

Area $A$ is $\pm \begin{vmatrix} a_{11} & a_{12} \\ a_{21} & a_{22} \end{vmatrix}$

18.4

The volume $V$ of the "box" spanned by the three vectors is
$$\pm \begin{vmatrix} a_{11} & a_{12} & a_{13} \\ a_{21} & a_{22} & a_{23} \\ a_{31} & a_{32} & a_{33} \end{vmatrix}$$

18.5 
$$\begin{vmatrix} a_{11} & a_{12} & \cdots & a_{1n} \\ \vdots & \vdots & & \vdots \\ a_{i1} & a_{i2} & \cdots & a_{in} \\ \vdots & \vdots & & \vdots \\ a_{n1} & a_{n2} & \cdots & a_{nn} \end{vmatrix} = a_{i1}A_{i1} + \cdots + a_{in}A_{in}$$

where $A_{ij}$, the *cofactor* of the element $a_{ij}$, is:

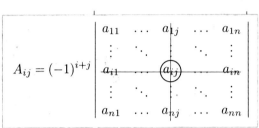

The general definition of a determinant of order $n$ in terms of cofactors. Its value is independent of the choice of $i$.

18.6 The determinant of **A** is zero if all elements in one row (or in one column) of **A** is 0.

Properties of determinants. **A** is a square matrix.

18.7 If two rows (or two columns) of **A** are interchanged, the determinant changes sign but the absolute value is unchanged.

18.8 A determinant is multiplied by a number when all of the elements in a single row (or column) of **A** are multiplied by that number.

18.9 If the elements in two rows (or two columns) of **A** are proportional, the determinant of **A** is 0.

18.10 The value of the determinant of **A** is unchanged if a multiple of one row (or one column) is added to another row (or column) of **A**.

18.11 If **A** is a square matrix, then the determinant of its transpose, **A**′, is, term by term, equal to the determinant of **A**: $|\mathbf{A}'| = |\mathbf{A}|$.

18.12 
$|\mathbf{A}\,\mathbf{B}| = |\mathbf{A}|\,|\mathbf{B}|$
$|\mathbf{A} + \mathbf{B}| \neq |\mathbf{A}| + |\mathbf{B}|$ in general

Properties of determinants. **A** and **B** are $n \times n$-matrices.

18.13 
$$\begin{vmatrix} 1 & x_1 & x_1^2 \\ 1 & x_2 & x_2^2 \\ 1 & x_3 & x_3^2 \end{vmatrix} = (x_2 - x_1)(x_3 - x_1)(x_3 - x_2)$$

The *Vandermonde determinant* for $n = 3$.

18.14 $\begin{vmatrix} 1 & x_1 & x_1^2 & \cdots & x_1^{n-1} \\ 1 & x_2 & x_2^2 & \cdots & x_2^{n-1} \\ \vdots & \vdots & \vdots & \ddots & \vdots \\ 1 & x_n & x_n^2 & \cdots & x_n^{n-1} \end{vmatrix} = \prod_{1 \leq j < i \leq n} (x_i - x_j)$    The general Vandermonde determinant.

18.15 $\begin{vmatrix} a_1 & 1 & 1 & \cdots & 1 \\ 1 & a_2 & 1 & \cdots & 1 \\ \vdots & \vdots & \vdots & \ddots & \vdots \\ 1 & 1 & 1 & \cdots & a_n \end{vmatrix}$
$= (a_1 - 1)(a_2 - 1) \cdots (a_n - 1)\left[1 + \sum_{i=1}^{n} \frac{1}{a_i - 1}\right]$    A special determinant.

18.16 $\begin{vmatrix} 0 & p_1 & \cdots & p_n \\ q_1 & a_{11} & \cdots & a_{1n} \\ \vdots & \vdots & \ddots & \vdots \\ q_n & a_{n1} & \cdots & a_{nn} \end{vmatrix} = -\sum_{i=1}^{n}\sum_{j=1}^{n} p_i A_{ji} q_j$    A useful determinant ($n \geq 2$). $A_{ji}$ is found in (18.5).

18.17 $\begin{vmatrix} \alpha & p_1 & \cdots & p_n \\ q_1 & a_{11} & \cdots & a_{1n} \\ \vdots & \vdots & \ddots & \vdots \\ q_n & a_{n1} & \cdots & a_{nn} \end{vmatrix} = (\alpha - \mathbf{P}'\mathbf{A}^{-1}\mathbf{Q})|\mathbf{A}|$    Generalization of (18.16) when $\mathbf{A}^{-1}$ exists. $\mathbf{P}' = (p_1, \ldots, p_n)$, $\mathbf{Q}' = (q_1, \ldots, q_n)$.

18.18 $|\mathbf{AB} + \mathbf{I}_n| = |\mathbf{BA} + \mathbf{I}_m|$    A useful result. $\mathbf{A}$ is $n \times m$, $\mathbf{B}$ is $m \times n$.

18.19 $\mathbf{A}\begin{pmatrix} i_1 & i_2 & \cdots & i_p \\ k_1 & k_2 & \cdots & k_p \end{pmatrix} = \begin{vmatrix} a_{i_1 k_1} & a_{i_1 k_2} & \cdots & a_{i_1 k_p} \\ a_{i_2 k_1} & a_{i_2 k_2} & \cdots & a_{i_2 k_p} \\ \vdots & \vdots & \ddots & \vdots \\ a_{i_p k_1} & a_{i_p k_2} & \cdots & a_{i_p k_p} \end{vmatrix}$    Notation for the determinant formed by rows $i_1, i_2, \ldots, i_p$, and columns $k_1, k_2, \ldots, k_p$, of $\mathbf{A} = (a_{ij})_{m \times n}$, $p \leq \min\{m, n\}$.

18.20 $\mathbf{A}\begin{pmatrix} i_1 & i_2 & \cdots & i_p \\ k_1 & k_2 & \cdots & k_p \end{pmatrix}$    *Minors* of $\mathbf{A} = (a_{ij})_{m \times n}$, if $1 \leq i_1 < i_2 < \cdots < i_p \leq n$ $1 \leq k_1 < k_2 < \cdots < k_p \leq n$.

18.21 $\mathbf{A}\begin{pmatrix} i_1 & i_2 & \cdots & i_p \\ i_1 & i_2 & \cdots & i_p \end{pmatrix}$ | *Principal minors* of $\mathbf{A} = (a_{ij})_{n \times n}$.

18.22 $\mathbf{A}\begin{pmatrix} 1 & 2 & \cdots & p \\ 1 & 2 & \cdots & p \end{pmatrix} = \begin{vmatrix} a_{11} & a_{12} & \cdots & a_{1p} \\ a_{21} & a_{22} & \cdots & a_{2p} \\ \vdots & \vdots & \ddots & \vdots \\ a_{p1} & a_{p2} & \cdots & a_{pp} \end{vmatrix}$ | *Leading* principal minors of $\mathbf{A} = (a_{ij})_{n \times n}$, consisting of the first $p$ rows and columns of $\mathbf{A}$.

If $|\mathbf{A}| = |(a_{ij})| \neq 0$, then the linear system of $n$ equations and $n$ unknowns,

$$a_{11}x_1 + a_{12}x_2 + \cdots + a_{1n}x_n = b_1$$
$$a_{21}x_1 + a_{22}x_2 + \cdots + a_{2n}x_n = b_2$$
$$\cdots\cdots\cdots\cdots\cdots\cdots\cdots\cdots\cdots\cdots$$
$$a_{n1}x_1 + a_{n2}x_2 + \cdots + a_{nn}x_n = b_n$$

has a unique solution

18.23
$$x_j = \frac{|\mathbf{A}_j|}{|\mathbf{A}|}, \quad j = 1, 2, \ldots, n$$
| *Cramer's rule.*

where

$$|\mathbf{A}_j| = \begin{vmatrix} a_{11} & \cdots & a_{1j-1} & b_1 & a_{1j+1} & \cdots & a_{1n} \\ a_{21} & \cdots & a_{2j-1} & b_2 & a_{2j+1} & \cdots & a_{2n} \\ \vdots & \ddots & \vdots & \vdots & \vdots & \ddots & \vdots \\ a_{n1} & \cdots & a_{nj-1} & b_n & a_{nj+1} & \cdots & a_{nn} \end{vmatrix}$$

## References

Most of the formulas are standard and are found in almost any linear algebra text, e.g. Anton (1987) or Lang (1987). A standard reference containing it all is Gantmacher (1959).

# Chapter 19

# Matrices

19.1 $\quad \mathbf{A} = \begin{pmatrix} a_{11} & a_{12} & \cdots & a_{1n} \\ a_{21} & a_{22} & \cdots & a_{2n} \\ \vdots & \vdots & & \vdots \\ a_{m1} & a_{m2} & \cdots & a_{mn} \end{pmatrix} = (a_{ij})_{m \times n}$ 

Notation for a *matrix*, where $a_{ij}$ is the element in the $i$th row and the $j$th column. The matrix has *order* $m \times n$. If $m = n$, the matrix is *square* of order $n$.

19.2 $\quad \mathrm{diag}(a_1, a_2, \ldots, a_n) = \begin{pmatrix} a_1 & 0 & \cdots & 0 \\ 0 & a_2 & \cdots & 0 \\ \vdots & \vdots & \ddots & \vdots \\ 0 & 0 & \cdots & a_n \end{pmatrix}$

A *diagonal matrix*.

19.3 $\quad \begin{pmatrix} a & 0 & \cdots & 0 \\ 0 & a & \cdots & 0 \\ \vdots & \vdots & \ddots & \vdots \\ 0 & 0 & \cdots & a \end{pmatrix}_{n \times n}$

A *scalar* matrix.

19.4 $\quad \mathbf{I}_n = \begin{pmatrix} 1 & 0 & \cdots & 0 \\ 0 & 1 & \cdots & 0 \\ \vdots & \vdots & \ddots & \vdots \\ 0 & 0 & \cdots & 1 \end{pmatrix}_{n \times n}$

The *unit* matrix or the *identity* matrix.

If $\mathbf{A} = (a_{ij})_{m \times n}$, $\mathbf{B} = (b_{ij})_{m \times n}$ and $\alpha$ is a scalar, we define

19.5 $\quad \mathbf{A} + \mathbf{B} = (a_{ij} + b_{ij})_{m \times n}$
$\quad \alpha \mathbf{A} = (\alpha a_{ij})_{m \times n}$
$\quad \mathbf{A} - \mathbf{B} = \mathbf{A} + (-1)\mathbf{B} = (a_{ij} - b_{ij})_{m \times n}$

Matrix operations. (The scalar is a real or complex number).

19.6
$$(\mathbf{A}+\mathbf{B})+\mathbf{C} = \mathbf{A}+(\mathbf{B}+\mathbf{C})$$
$$\mathbf{A}+\mathbf{B} = \mathbf{B}+\mathbf{A}$$
$$\mathbf{A}+\mathbf{0} = \mathbf{A}$$
$$\mathbf{A}+(-\mathbf{A}) = \mathbf{0}$$
$$(a+b)\mathbf{A} = a\mathbf{A}+b\mathbf{A}$$
$$a(\mathbf{A}+\mathbf{B}) = a\mathbf{A}+a\mathbf{B}$$

Properties of matrix operations. $\mathbf{0}$ is the zero (or null) matrix, all of whose elements are zero. $a$ and $b$ are scalars.

19.7 If $\mathbf{A} = (a_{ij})_{m \times n}$ and $\mathbf{B} = (b_{ij})_{n \times p}$, we define the *product* $\mathbf{C} = \mathbf{AB}$ as that $m \times p$-matrix $\mathbf{C} = (c_{ij})_{m \times p}$ where
$$c_{ij} = a_{i1}b_{1j} + \cdots + a_{ik}b_{kj} + \cdots + a_{in}b_{nj}$$

The definition of *matrix multiplication*.

$$\begin{pmatrix} a_{11} & \cdots & a_{1k} & \cdots & a_{1n} \\ \vdots & & \vdots & & \vdots \\ \boxed{a_{i1} & \cdots & a_{ik} & \cdots & a_{in}} \\ \vdots & & \vdots & & \vdots \\ a_{m1} & \cdots & a_{mk} & \cdots & a_{mn} \end{pmatrix} \cdot \begin{pmatrix} b_{11} & \cdots & \boxed{b_{1j}} & \cdots & b_{1p} \\ \vdots & & \vdots & & \vdots \\ b_{k1} & \cdots & \boxed{b_{kj}} & \cdots & b_{kp} \\ \vdots & & \vdots & & \vdots \\ b_{n1} & \cdots & \boxed{b_{nj}} & \cdots & b_{np} \end{pmatrix} = \begin{pmatrix} c_{11} & \cdots & c_{1j} & \cdots & c_{1p} \\ \vdots & & \vdots & & \vdots \\ c_{i1} & \cdots & \boxed{c_{ij}} & \cdots & c_{ip} \\ \vdots & & \vdots & & \vdots \\ c_{m1} & \cdots & c_{mj} & \cdots & c_{mp} \end{pmatrix}$$

19.8
$$(\mathbf{AB})\mathbf{C} = \mathbf{A}(\mathbf{BC})$$
$$\mathbf{A}(\mathbf{B}+\mathbf{C}) = \mathbf{AB}+\mathbf{AC}$$
$$(\mathbf{A}+\mathbf{B})\mathbf{C} = \mathbf{AC}+\mathbf{BC}$$

Properties of matrix multiplication.

19.9
$$\mathbf{AB} \neq \mathbf{BA}$$
$$\mathbf{AB} = \mathbf{0} \not\Rightarrow \mathbf{A} = \mathbf{0} \ \text{ or } \ \mathbf{B} = \mathbf{0}$$
$$\mathbf{AB} = \mathbf{AC} \ \& \ \mathbf{A} \neq \mathbf{0} \not\Rightarrow \mathbf{B} = \mathbf{C}$$

Important facts about matrix multiplication. $\mathbf{0}$ is the zero matrix.

19.10
$$\mathbf{A}' = \begin{pmatrix} a_{11} & a_{21} & \cdots & a_{m1} \\ a_{12} & a_{22} & \cdots & a_{m2} \\ \vdots & \vdots & & \vdots \\ a_{1n} & a_{2n} & \cdots & a_{mn} \end{pmatrix}$$

$\mathbf{A}'$, the *transpose* of $\mathbf{A} = (a_{ij})_{m \times n}$, is the $n \times m$-matrix obtained by interchanging rows and columns in $\mathbf{A}$.

19.11
$$(\mathbf{A}')' = \mathbf{A}$$
$$(\mathbf{A}+\mathbf{B})' = \mathbf{A}'+\mathbf{B}'$$
$$(\alpha\mathbf{A})' = \alpha\mathbf{A}'$$
$$(\mathbf{AB})' = \mathbf{B}'\mathbf{A}' \ \text{(NOTE the order!)}$$

Rules for transposes.

19.12 A square matrix $\mathbf{A}$ of order $n$ is called
- *symmetric* if $\mathbf{A} = \mathbf{A}'$
- *anti-symmetric* if $\mathbf{A} = -\mathbf{A}'$
- *idempotent* if $\mathbf{A}^2 = \mathbf{A}$
- *involutive* if $\mathbf{A}^2 = \mathbf{I}_n$
- *orthogonal* if $\mathbf{A}'\mathbf{A} = \mathbf{I}_n$

Some special matrices. (See also Chapter 21.)

19.13 $\quad \mathbf{B} = \mathbf{A}^{-1} \iff \mathbf{AB} = \mathbf{I}_n \quad \text{or} \quad \mathbf{BA} = \mathbf{I}_n$

The *inverse* of an $n \times n$-matrix $\mathbf{A}$. $\mathbf{I}_n$ is the identity matrix.

19.14 $\quad \mathbf{A}^{-1} \text{ exists} \iff |\mathbf{A}| \neq 0$

A necessary and sufficient condition for a matrix to have an inverse, i.e. be *invertible*.

19.15 $\quad \mathbf{A} = \begin{pmatrix} a & b \\ c & d \end{pmatrix} \implies \mathbf{A}^{-1} = \dfrac{1}{ad - bc} \begin{pmatrix} d & -b \\ -c & a \end{pmatrix}$

Valid if $|\mathbf{A}| = ad - bc \neq 0$.

19.16 If $\mathbf{A} = \begin{pmatrix} a_{11} & a_{12} & \cdots & a_{1n} \\ a_{21} & a_{22} & \cdots & a_{2n} \\ \vdots & \vdots & \ddots & \vdots \\ a_{n1} & a_{n2} & \cdots & a_{nn} \end{pmatrix}$ and $|\mathbf{A}| \neq 0$, then

$$\mathbf{A}^{-1} = \frac{1}{|\mathbf{A}|} \begin{pmatrix} A_{11} & A_{21} & \cdots & A_{n1} \\ A_{12} & A_{22} & \cdots & A_{n2} \\ \vdots & \vdots & \ddots & \vdots \\ A_{1n} & A_{2n} & \cdots & A_{nn} \end{pmatrix}$$

where the *cofactor*, $A_{ij}$, of the element $a_{ij}$ is

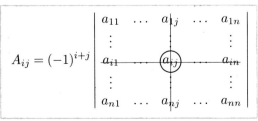

The general formula for the inverse of a square matrix. (NOTE the order of the indices in the cofactor matrix!)

19.17
$$(\mathbf{A}^{-1})^{-1} = \mathbf{A}$$
$$(\mathbf{AB})^{-1} = \mathbf{B}^{-1}\mathbf{A}^{-1} \quad \text{(NOTE the order!)}$$
$$(\mathbf{A}')^{-1} = (\mathbf{A}^{-1})'$$
$$(c\mathbf{A})^{-1} = c^{-1}\mathbf{A}^{-1}$$

Properties of the inverse. ($\mathbf{A}$ and $\mathbf{B}$ are invertible $n \times n$-matrices. $c$ is a scalar $\neq 0$.)

19.18 $(\mathbf{I}_m + \mathbf{AB})^{-1} = \mathbf{I}_m - \mathbf{A}(\mathbf{I}_n + \mathbf{BA})^{-1}\mathbf{B}$

$\mathbf{A}$ is $m \times n$, $\mathbf{B}$ is $n \times m$, $|\mathbf{I}_m + \mathbf{AB}| \neq 0$.

19.19 $\mathbf{R}^{-1}\mathbf{A}'(\mathbf{A}\mathbf{R}^{-1}\mathbf{A}' + \mathbf{Q}^{-1})^{-1} = (\mathbf{A}'\mathbf{Q}\mathbf{A} + \mathbf{R})^{-1}\mathbf{A}'\mathbf{Q}$

Matrix inversion pairs. Valid if the inverses exist.

19.20 $\operatorname{tr}(\mathbf{A}) = \sum_{i=1}^{n} a_{ii}$

The *trace* of a matrix $\mathbf{A} = (a_{ij})_{n \times n}$ is the sum of its diagonal elements.

19.21
$$\operatorname{tr}(\mathbf{A} + \mathbf{B}) = \operatorname{tr}(\mathbf{A}) + \operatorname{tr}(\mathbf{B})$$
$$\operatorname{tr}(c\mathbf{A}) = c\operatorname{tr}(\mathbf{A}) \quad (c \text{ is a scalar})$$
$$\operatorname{tr}(\mathbf{AB}) = \operatorname{tr}(\mathbf{BA}) \quad (\text{whenever } \mathbf{AB} \text{ is square})$$
$$\operatorname{tr}(\mathbf{A}') = \operatorname{tr}(\mathbf{A})$$

Properties of the trace.

19.22 $r(\mathbf{A}) =$ max. number of lin. indep. rows in $\mathbf{A}$
$=$ max. number of lin. indep. columns in $\mathbf{A}$
$=$ order of the largest nonzero minor of $\mathbf{A}$.

Definitions of the *rank* of a matrix $\mathbf{A}$. On minors, see (18.20).

19.23
$$r(\mathbf{A}) = r(\mathbf{A}') = r(\mathbf{A}'\mathbf{A}) = r(\mathbf{A}\mathbf{A}')$$
$$r(\mathbf{A}\,\mathbf{B}) \leq \min\bigl(r(\mathbf{A}), r(\mathbf{B})\bigr)$$
$$r(\mathbf{AB}) = r(\mathbf{B}), \quad (|\mathbf{A}| \neq 0)$$
$$r(\mathbf{CA}) = r(\mathbf{C}), \quad (|\mathbf{A}| \neq 0)$$
$$r(\mathbf{PAQ}) = r(\mathbf{A}), \quad (|\mathbf{P}| \neq 0, |\mathbf{Q}| \neq 0)$$
$$|r(\mathbf{A}) - r(\mathbf{B})| \leq r(\mathbf{A} + \mathbf{B}) \leq r(\mathbf{A}) + r(\mathbf{B})$$
$$r(\mathbf{AB}) \geq r(\mathbf{A}) + r(\mathbf{B}) - n$$
$$r(\mathbf{AB}) + r(\mathbf{BC}) \leq r(\mathbf{B}) + r(\mathbf{ABC})$$

Properties of the rank. The order of the matrices are such that the required operations are defined. In result no 7 (*Sylvesters' inequality*), $\mathbf{A}$ and $\mathbf{B}$ are $n \times n$. Result no 8 is called *Frobenius' inequality*.

19.24 $\mathbf{Ax} = \mathbf{0}$ for some $\mathbf{x} \neq \mathbf{0} \iff r(\mathbf{A}) \leq n - 1$

Useful result on homogeneous equations. $\mathbf{A}$ is $m \times n$, $\mathbf{x}$ is $n \times 1$.

| | | |
|---|---|---|
| 19.25 | A norm for a square matrix $\mathbf{A}$ is a real number, $\|\mathbf{A}\|_\beta$, such that:<br>• $\|\mathbf{A}\|_\beta > 0$ for $\mathbf{A} \neq \mathbf{0}$ and $\|\mathbf{0}\|_\beta = 0$<br>• $\|c\mathbf{A}\|_\beta = |c|\,\|\mathbf{A}\|_\beta$ ($c$ a scalar)<br>• $\|\mathbf{A}+\mathbf{B}\|_\beta \leq \|\mathbf{A}\|_\beta + \|\mathbf{B}\|_\beta$<br>• $\|\mathbf{A}\,\mathbf{B}\|_\beta \leq \|\mathbf{A}\|_\beta\,\|\mathbf{B}\|_\beta$ | Definition of *matrix norm*. (There are an infinite number of such norms, some of which are given in (19.26).) |
| 19.26 | • $\|\mathbf{A}\|_1 = \max\limits_{j=1,\ldots,n} \sum\limits_{i=1}^{n} |a_{ij}|$<br>• $\|\mathbf{A}\|_\infty = \max\limits_{i=1,\ldots,n} \sum\limits_{j=1}^{n} |a_{ij}|$<br>• $\|\mathbf{A}\|_2 = \sqrt{\lambda}$, where $\lambda$ is the largest characteristic root of $\mathbf{A}'\mathbf{A}$.<br>• $\|\mathbf{A}\|_M = n \max\limits_{i,j=1,\ldots,n} |a_{ij}|$<br>• $\|\mathbf{A}\|_T = \left( \sum\limits_{j=1}^{n} \sum\limits_{i=1}^{n} |a_{ij}|^2 \right)^{1/2}$ | Some matrix norms for $\mathbf{A} = (a_{ij})_{n\times n}$. (For characteristic roots see Chapter 20.) |
| 19.27 | $\lambda$ char. root for $\mathbf{A} = (a_{ij})_{n\times n} \Rightarrow |\lambda| \leq \|\mathbf{A}\|_\beta$ | The modulus of any characteristic root of $\mathbf{A}$ is less than or equal to any matrix norm for $\mathbf{A}$. |
| 19.28 | $\|\mathbf{A}\|_\beta < 1 \Rightarrow \mathbf{A}^t \to \mathbf{0}$ as $t \to \infty$ | Sufficient condition for $\mathbf{A}^t \to \mathbf{0}$ as $t \to \infty$. $\|\mathbf{A}\|_\beta$ is any matrix norm of $\mathbf{A}$. |
| 19.29 | $\mathbf{P} = \begin{pmatrix} \mathbf{P}_{11} & \mathbf{P}_{12} \\ \mathbf{P}_{21} & \mathbf{P}_{22} \end{pmatrix}$ | A *partitioned* matrix of order $(p+q) \times (r+s)$. ($\mathbf{P}_{11}$ is $p\times r$, $\mathbf{P}_{12}$ is $p\times s$, $\mathbf{P}_{21}$ is $q\times r$, $\mathbf{P}_{22}$ is $q\times s$.) |
| 19.30 | $\mathbf{P} = \begin{pmatrix} \mathbf{P}_{11} & \mathbf{P}_{12} \\ \mathbf{P}_{21} & \mathbf{P}_{22} \end{pmatrix}, \quad \mathbf{Q} = \begin{pmatrix} \mathbf{Q}_{11} & \mathbf{Q}_{12} \\ \mathbf{Q}_{21} & \mathbf{Q}_{22} \end{pmatrix} \Rightarrow$<br>$\mathbf{PQ} = \begin{pmatrix} \mathbf{P}_{11}\mathbf{Q}_{11}+\mathbf{P}_{12}\mathbf{Q}_{21} & \mathbf{P}_{11}\mathbf{Q}_{12}+\mathbf{P}_{12}\mathbf{Q}_{22} \\ \mathbf{P}_{21}\mathbf{Q}_{11}+\mathbf{P}_{22}\mathbf{Q}_{21} & \mathbf{P}_{21}\mathbf{Q}_{12}+\mathbf{P}_{22}\mathbf{Q}_{22} \end{pmatrix}$ | Multiplication of partitioned matrices. (We assume that the multiplications involved are defined.) |
| 19.31 | $\mathbf{P} = \begin{pmatrix} \mathbf{P}_{11} & \mathbf{P}_{12} \\ \mathbf{P}_{21} & \mathbf{P}_{22} \end{pmatrix} \Rightarrow$<br>$|\mathbf{P}| = |\mathbf{P}_{11}| \cdot |\mathbf{P}_{22} - \mathbf{P}_{21}\mathbf{P}_{11}^{-1}\mathbf{P}_{12}|$ | The determinant of a partitioned $n\times n$-matrix, assuming $\mathbf{P}_{11}^{-1}$ exists. |

19.32  $\mathbf{P} = \begin{pmatrix} \mathbf{P}_{11} & \mathbf{P}_{12} \\ \mathbf{P}_{21} & \mathbf{P}_{22} \end{pmatrix} \Rightarrow$

$|\mathbf{P}| = |\mathbf{P}_{22}| \cdot |\mathbf{P}_{11} - \mathbf{P}_{12}\mathbf{P}_{22}^{-1}\mathbf{P}_{21}|$

The determinant of a partitioned $n \times n$-matrix, assuming $\mathbf{P}_{22}^{-1}$ exists.

19.33  $\mathbf{P} = \begin{pmatrix} \mathbf{P}_{11} & \mathbf{0} \\ \mathbf{P}_{21} & \mathbf{P}_{22} \end{pmatrix} \Rightarrow |\mathbf{P}| = |\mathbf{P}_{11}| \cdot |\mathbf{P}_{22}|$

A special case.

19.34  $\mathbf{P} = \begin{pmatrix} \mathbf{P}_{11} & \mathbf{P}_{12} \\ \mathbf{0} & \mathbf{P}_{22} \end{pmatrix} \Rightarrow |\mathbf{P}| = |\mathbf{P}_{11}| \cdot |\mathbf{P}_{22}|$

A special case.

19.35  $\begin{pmatrix} \mathbf{P}_{11} & \mathbf{P}_{12} \\ \mathbf{P}_{21} & \mathbf{P}_{22} \end{pmatrix}^{-1} =$

$\begin{pmatrix} \mathbf{P}_{11}^{-1} + \mathbf{P}_{11}^{-1}\mathbf{P}_{12}\boldsymbol{\Delta}^{-1}\mathbf{P}_{21}\mathbf{P}_{11}^{-1} & -\mathbf{P}_{11}^{-1}\mathbf{P}_{12}\boldsymbol{\Delta}^{-1} \\ -\boldsymbol{\Delta}^{-1}\mathbf{P}_{21}\mathbf{P}_{11}^{-1} & \boldsymbol{\Delta}^{-1} \end{pmatrix}$

where $\boldsymbol{\Delta} = \mathbf{P}_{22} - \mathbf{P}_{21}\mathbf{P}_{11}^{-1}\mathbf{P}_{12}$

The inverse of a partitioned matrix, assuming $\mathbf{P}_{11}^{-1}$ exists.

19.36  $\begin{pmatrix} \mathbf{P}_{11} & \mathbf{P}_{12} \\ \mathbf{P}_{21} & \mathbf{P}_{22} \end{pmatrix}^{-1} =$

$\begin{pmatrix} \boldsymbol{\Delta}_1^{-1} & -\boldsymbol{\Delta}_1^{-1}\mathbf{P}_{12}\mathbf{P}_{22}^{-1} \\ -\mathbf{P}_{22}^{-1}\mathbf{P}_{21}\boldsymbol{\Delta}_1^{-1} & \mathbf{P}_{22}^{-1} + \mathbf{P}_{22}^{-1}\mathbf{P}_{21}\boldsymbol{\Delta}_1^{-1}\mathbf{P}_{12}\mathbf{P}_{22}^{-1} \end{pmatrix}$

where $\boldsymbol{\Delta}_1 = \mathbf{P}_{11} - \mathbf{P}_{12}\mathbf{P}_{22}^{-1}\mathbf{P}_{21}$

The inverse of a partitioned matrix, assuming $\mathbf{P}_{22}^{-1}$ exists.

## References

Most of the formulas are standard and are found in almost any linear algebra text, e.g. Anton (1987) or Lang (1987). For (19.25)–(19.28) see e.g. Faddeeva (1959). A standard reference containing it all is Gantmacher (1959).

# Chapter 20

# Characteristic roots. Quadratic forms

| | | |
|---|---|---|
| 20.1 | $\mathbf{Ac} = \lambda \mathbf{c}$ | $\lambda$ is a *characteristic root* (an *eigenvalue*) and $\mathbf{c} \neq \mathbf{0}$ is a *characteristic vector* (an *eigenvector*) of an $n \times n$-matrix $\mathbf{A}$. |
| 20.2 | $\|\mathbf{A} - \lambda\mathbf{I}\| = \begin{vmatrix} a_{11}-\lambda & a_{12} & \cdots & a_{1n} \\ a_{21} & a_{22}-\lambda & \cdots & a_{2n} \\ \vdots & \vdots & \ddots & \vdots \\ a_{n1} & a_{n2} & \cdots & a_{nn}-\lambda \end{vmatrix}$ | The *characteristic polynomial* of $\mathbf{A} = (a_{ij})_{n \times n}$. $\mathbf{I}$ is the unit matrix of order $n$. |
| 20.3 | $\lambda$ is a char. root of $\mathbf{A} \Leftrightarrow p(\lambda) = \|\mathbf{A} - \lambda\mathbf{I}\| = 0$ | A necessary and sufficient condition for $\lambda$ to be a characteristic root of $\mathbf{A}$. |
| 20.4 | $\begin{vmatrix} a_{11}-\lambda & a_{12} \\ a_{21} & a_{22}-\lambda \end{vmatrix} = \lambda^2 - (\operatorname{tr}(\mathbf{A}))\lambda + \|\mathbf{A}\|$ | The characteristic polynomial of $\mathbf{A} = (a_{ij})_{2 \times 2}$. (For $\operatorname{tr}(\mathbf{A})$ see (19.20).) |
| 20.5 | $\|\mathbf{A}\| = \lambda_1 \cdot \lambda_2 \cdots \lambda_{n-1} \cdot \lambda_n$ <br> $\operatorname{tr}(\mathbf{A}) = a_{11} + \cdots + a_{nn} = \lambda_1 + \cdots + \lambda_n$ | $\lambda_1, \ldots, \lambda_n$ are the characteristic roots of $\mathbf{A}$. |
| 20.6 | Let $f$ be a polynomial. If $\lambda_1, \ldots, \lambda_n$ are characteristic roots of $\mathbf{A}$, then $f(\lambda_1), \ldots, f(\lambda_n)$ are characteristic roots of $f(\mathbf{A})$. | Characteristic roots of a polynomial of a matrix. |
| 20.7 | If $\lambda_1, \ldots, \lambda_n$ are characteristic roots of $\mathbf{A}$, then $\lambda_1^{-1}, \ldots, \lambda_n^{-1}$ are characteristic roots of $\mathbf{A}^{-1}$. | Valid if $\mathbf{A}^{-1}$ exists. |
| 20.8 | All characteristic roots of $\mathbf{A}$ have moduli (strictly) less than 1, if and only if $\mathbf{A}^t \to \mathbf{0}$ as $t \to \infty$. | An important result. |

| | | |
|---|---|---|
| 20.9 | **AB** and **BA** have the same characteristic roots. | **A** and **B** are $n \times n$-matrices. |
| 20.10 | If **A** is symmetric and has real numbers as elements, all characteristic roots of **A** are real numbers. | An important result. |
| 20.11 | **A** is *diagonalizable* $\Leftrightarrow$ $\begin{cases} \mathbf{P}^{-1}\mathbf{AP} = \mathbf{D} \text{ for a} \\ \text{matrix } \mathbf{P} \text{ and some} \\ \text{diagonal matrix } \mathbf{D}. \end{cases}$ | A definition. |
| 20.12 | **A** and $\mathbf{P}^{-1}\mathbf{AP}$ have the same char. roots. | A simple fact. |
| 20.13 | If $\mathbf{A} = (a_{ij})_{n \times n}$ has $n$ distinct characteristic roots, then **A** is diagonalizable. | Sufficient (but NOT necessary) condition for **A** to be diagonalizable. |
| 20.14 | Given any matrix $\mathbf{A} = (a_{ij})_{n \times n}$, there is for every $\epsilon > 0$ a matrix $\mathbf{B}_\epsilon = (b_{ij})_{n \times n}$, with *distinct* characteristic roots, such that $$\sum_{i,j=1}^{n} |a_{ij} - b_{ij}| < \epsilon$$ | By changing the elements of a matrix only slightly you can get a matrix with distinct characteristic roots. |
| 20.15 | $\mathbf{A} = (a_{ij})_{n \times n}$ has $n$ linearly independent characteristic column vectors, $\mathbf{x}_1, \ldots, \mathbf{x}_n$, with associated characteristic roots $\lambda_1, \ldots, \lambda_n$ if and only if $$\mathbf{P}^{-1}\mathbf{AP} = \begin{pmatrix} \lambda_1 & 0 & \cdots & 0 \\ 0 & \lambda_2 & \cdots & 0 \\ \vdots & \vdots & \ddots & \vdots \\ 0 & 0 & \cdots & \lambda_n \end{pmatrix}$$ where $\mathbf{P} = (\mathbf{x}_1, \ldots, \mathbf{x}_n)_{n \times n}$. | A characterization of diagonalizable matrices. |
| 20.16 | If $\mathbf{A} = (a_{ij})_{n \times n}$ is symmetric and $\lambda_1, \lambda_2, \ldots, \lambda_n$ are its characteristic roots, there exists an orthogonal matrix **U** such that $$\mathbf{U}^{-1}\mathbf{AU} = \begin{pmatrix} \lambda_1 & 0 & \cdots & 0 \\ 0 & \lambda_2 & \cdots & 0 \\ \vdots & \vdots & \ddots & \vdots \\ 0 & 0 & \cdots & \lambda_n \end{pmatrix}$$ | The *spectral theorem* for symmetric matrices. For properties of orthogonal matrices, see Chapter 21. |

20.17 If **A** is an $n \times n$-matrix with characteristic roots $\lambda_1, \ldots, \lambda_n$ (not necessarily distinct), then there exists an invertible $n \times n$-matrix **T** such that

$$\mathbf{T}^{-1}\mathbf{A}\mathbf{T} = \begin{pmatrix} \mathbf{J}_{k_1}(\lambda_1) & 0 & \cdots & 0 \\ 0 & \mathbf{J}_{k_2}(\lambda_2) & \cdots & 0 \\ \vdots & \vdots & \ddots & \vdots \\ 0 & 0 & \cdots & \mathbf{J}_{k_r}(\lambda_r) \end{pmatrix}$$

where $k_1 + k_2 + \cdots + k_r = n$ and $\mathbf{J}_k$ is the $k \times k$-matrix

$$\mathbf{J}_k(\lambda) = \begin{pmatrix} \lambda & 1 & 0 & \cdots & 0 \\ 0 & \lambda & 1 & \cdots & 0 \\ \vdots & \vdots & \vdots & \ddots & \vdots \\ 0 & 0 & 0 & \cdots & 1 \\ 0 & 0 & 0 & \cdots & \lambda \end{pmatrix}, \quad \mathbf{J}_1(\lambda) = \lambda$$

The *Jordan decomposition theorem.*

20.18 A square matrix **A** satisfies its own characteristic equation.

The *Cayley-Hamilton* theorem.

20.19 $\mathbf{A} = \begin{pmatrix} a_{11} & a_{12} \\ a_{21} & a_{22} \end{pmatrix} \Rightarrow \mathbf{A}^2 - (\operatorname{tr}(\mathbf{A}))\mathbf{A} + |\mathbf{A}|\mathbf{I} = \mathbf{0}$

The *Cayley-Hamilton* theorem for $n = 2$. (See (20.4).)

20.20 $Q = \sum_{i=1}^{n} \sum_{j=1}^{n} a_{ij} x_i x_j =$
$a_{11} x_1^2 + a_{12} x_1 x_2 + \cdots + a_{ij} x_i x_j + \cdots + a_{nn} x_n^2$

A *quadratic form* in $n$ variables $x_1, \ldots, x_n$. We can assume, without loss of generality, that $a_{ij} = a_{ji}$ for all $i, j = 1, \ldots, n$.

20.21 $\mathbf{x} = \begin{pmatrix} x_1 \\ \vdots \\ x_n \end{pmatrix}, \quad \mathbf{A} = \begin{pmatrix} a_{11} & a_{12} & \cdots & a_{1n} \\ a_{21} & a_{22} & \cdots & a_{2n} \\ \vdots & \vdots & \ddots & \vdots \\ a_{n1} & a_{n2} & \cdots & a_{nn} \end{pmatrix}$

implies

$Q = \sum_{i=1}^{n} \sum_{j=1}^{n} a_{ij} x_i x_j = \mathbf{x}' \mathbf{A} \mathbf{x}$

A quadratic form in matrix formulation. We can assume, without loss of generality, that **A** is symmetric.

| | | |
|---|---|---|
| 20.22 | $\mathbf{x}'\mathbf{A}\mathbf{x}$ is PD $\Leftrightarrow \mathbf{x}'\mathbf{A}\mathbf{x} > 0$ for all $\mathbf{x} \neq \mathbf{0}$<br>$\mathbf{x}'\mathbf{A}\mathbf{x}$ is PSD $\Leftrightarrow \mathbf{x}'\mathbf{A}\mathbf{x} \geq 0$ for all $\mathbf{x}$<br>$\mathbf{x}'\mathbf{A}\mathbf{x}$ is ND $\Leftrightarrow \mathbf{x}'\mathbf{A}\mathbf{x} < 0$ for all $\mathbf{x} \neq \mathbf{0}$<br>$\mathbf{x}'\mathbf{A}\mathbf{x}$ is NSD $\Leftrightarrow \mathbf{x}'\mathbf{A}\mathbf{x} \leq 0$ for all $\mathbf{x}$<br>$\mathbf{x}'\mathbf{A}\mathbf{x}$ is ID $\Leftrightarrow \mathbf{x}'\mathbf{A}\mathbf{x}$ is neither PSD, nor NSD | Definition of *positive definite* (PD), *positive semidefinite* (PSD), *negative definite* (ND), *negative semidefinite* (NSD), and *indefinite* (ID) quadratic forms (matrices). |
| 20.23 | $\mathbf{x}'\mathbf{A}\mathbf{x}$ is PD $\Rightarrow a_{ii} > 0$ for $i = 1, \ldots, n$<br>$\mathbf{x}'\mathbf{A}\mathbf{x}$ is PSD $\Rightarrow a_{ii} \geq 0$ for $i = 1, \ldots, n$<br>$\mathbf{x}'\mathbf{A}\mathbf{x}$ is ND $\Rightarrow a_{ii} < 0$ for $i = 1, \ldots, n$<br>$\mathbf{x}'\mathbf{A}\mathbf{x}$ is NSD $\Rightarrow a_{ii} \leq 0$ for $i = 1, \ldots, n$ | Follows from (20.22) by letting $x_j = 0$ for $j \neq i$, $x_i = 1$. |
| 20.24 | $\mathbf{x}'\mathbf{A}\mathbf{x}$ is PD $\Leftrightarrow$ all the char. roots of $\mathbf{A}$ are $> 0$<br>$\mathbf{x}'\mathbf{A}\mathbf{x}$ is PSD $\Leftrightarrow$ all the char. roots of $\mathbf{A}$ are $\geq 0$<br>$\mathbf{x}'\mathbf{A}\mathbf{x}$ is ND $\Leftrightarrow$ all the char. roots of $\mathbf{A}$ are $< 0$<br>$\mathbf{x}'\mathbf{A}\mathbf{x}$ is NSD $\Leftrightarrow$ all the char. roots of $\mathbf{A}$ are $\leq 0$ | A characterization of definite quadratic forms (matrices) in terms of characteristic roots. |
| 20.25 | $\mathbf{x}'\mathbf{A}\mathbf{x}$ is indefinite (ID) if and only if there exist at least one positive and one negative characteristic root of $\mathbf{A}$. | A characterization of *indefinite forms*. |
| 20.26 | $D_k = \begin{vmatrix} a_{11} & a_{12} & \cdots & a_{1k} \\ a_{21} & a_{22} & \cdots & a_{2k} \\ \vdots & \vdots & \ddots & \vdots \\ a_{k1} & a_{k2} & \cdots & a_{kk} \end{vmatrix}$, $k = 1, 2, \ldots, n$ | The leading principal minors of $\mathbf{A} = (a_{ij})_{n \times n}$. (See (18.22).) |
| 20.27 | $\mathbf{x}'\mathbf{A}\mathbf{x}$ is PD $\Leftrightarrow D_k > 0$ for all $k = 1, \ldots, n$<br>$\mathbf{x}'\mathbf{A}\mathbf{x}$ is ND $\Leftrightarrow (-1)^k D_k > 0$ for $k = 1, \ldots, n$. | A characterization of definite quadratic forms (matrices) in terms of leading principal minors. Note that replacing $>$ by $\geq$ will NOT give criteria for the semidefinite case. Example: $Q = 0x_1^2 + 0x_1x_2 - x_2^2$. |
| 20.28 | $\mathbf{x}'\mathbf{A}\mathbf{x}$ is PSD (NSD) $\Leftrightarrow \begin{cases} \text{All the principal} \\ \text{minors of } \mathbf{A} \text{ are} \\ \geq 0 \ (\leq 0) \end{cases}$ | Characterizations of positive and negative semidefinite quadratic forms (matrices) in terms of principal minors. (See (18.21).) |

| | | |
|---|---|---|
| 20.29 | $(*)$ $\quad Q = \sum_{i=1}^{n}\sum_{j=1}^{n} a_{ij}x_ix_j, \qquad (a_{ij}=a_{ji})$<br><br>is positive (negative) definite subject to the linear constraints<br><br>$(**)$ $\quad \begin{array}{c} b_{11}x_1 + \cdots + b_{1n}x_n = 0 \\ \cdots\cdots\cdots\cdots\cdots\cdots\cdots\cdots\cdots \\ b_{m1}x_1 + \cdots + b_{mn}x_n = 0 \end{array} \qquad (m < n)$<br><br>if $Q > 0\ (<0)$ for all $(x_1,\ldots,x_n) \neq (0,\ldots,0)$ satisfying $(**)$. | A definition of positive (negative) definiteness subject to linear constraints. |
| 20.30 | $D_r = \begin{vmatrix} 0 & \cdots & 0 & b_{11} & \cdots & b_{1r} \\ \vdots & & \vdots & \vdots & & \vdots \\ 0 & \cdots & 0 & b_{m1} & \cdots & b_{mr} \\ b_{11} & \cdots & b_{m1} & a_{11} & \cdots & a_{1r} \\ \vdots & & \vdots & \vdots & & \vdots \\ b_{1r} & \cdots & b_{mr} & a_{r1} & \cdots & a_{rr} \end{vmatrix}$ | A bordered determinant associated with (20.29), $r = 1,\ldots,n$. |
| 20.31 | Necessary and sufficient conditions for the quadratic form $(*)$ in (20.29) to be positive definite subject to the constraints $(**)$, assuming that the first $m$ columns of the matrix $(b_{ij})_{m\times n}$ are linearly independent, is that<br><br>$(-1)^m D_r > 0, \quad r = m+1,\ldots,n$<br><br>The corresponding conditions for $(*)$ to be negative definite subject to the constraints $(**)$ is that<br><br>$(-1)^r D_r > 0, \quad r = m+1,\ldots,n$ | A test for definiteness of quadratic forms subject to constraints. (Assuming that the rank of $(b_{ij})_{m\times n}$ is $m$ is not enough, as is shown by the example, $Q = x_1^2 + x_2^2 - x_3^2$ subject to $x_3 = 0$.) |
| 20.32 | $ax^2 + 2bxy + cy^2$ is positive for all $(x,y) \neq (0,0)$ where $px + qy = 0$, if and only if,<br><br>$\begin{vmatrix} 0 & p & q \\ p & a & b \\ q & b & c \end{vmatrix} < 0$ | A special case of (20.31), assuming $(p,q) \neq (0,0)$. |
| 20.33 | If $\mathbf{A} = (a_{ij})_{n\times n}$ is positive definite and $\mathbf{P}$ is $n\times m$ with $r(\mathbf{P}) = m$, then $\mathbf{P'AP}$ is positive definite. | |
| 20.34 | If $\mathbf{P}$ is $n\times m$ and $r(\mathbf{P}) = m$, then $\mathbf{PP'}$ is positive definite and has rank $m$. | Results on positive definite matrices. |

20.35 If **A** is positive definite, there exists a nonsingular matrix **P** such that $\mathbf{PAP'} = \mathbf{I}$ and $\mathbf{P'P} = \mathbf{A}^{-1}$.

20.36 Let **A** be an $n \times m$-matrix of rank $r$, $r \leq m \leq n$. Then there exist matrices $\mathbf{B}_1$, $\mathbf{B}_2$ and a diagonal matrix **D** with positive diagonal elements, such that $\mathbf{A} = \mathbf{B}_1 \mathbf{D} \mathbf{B}_2$. — The *singular value decomposition theorem*.

20.37 Let **A** and **B** be symmetric $n \times n$-matrices. Then there exists an orthogonal matrix **Q** such that $\mathbf{Q'AQ} = \mathbf{D}_1$ and $\mathbf{Q'BQ} = \mathbf{D}_2$, where $\mathbf{D}_1$ and $\mathbf{D}_2$ are diagonal matrices, if and only if, $\mathbf{AB} = \mathbf{BA}$. — *Simultaneous diagonalization*.

## References

Most of the formulas are found in almost any linear algebra text, e.g. Anton (1987) or Lang (1987). A standard reference containing it all is Gantmacher (1959).

# Chapter 21

# Special matrices. Leontief systems

## Idempotent matrices

| | | |
|---|---|---|
| 21.1 | $\mathbf{A} = (a_{ij})_{n \times n}$ is *idempotent* $\iff \mathbf{A}^2 = \mathbf{A}$ | Definition of an idempotent matrix. |
| 21.2 | $\mathbf{A}$ is idempotent $\iff \mathbf{I} - \mathbf{A}$ is idempotent. | Properties of idempotent matrices. |
| 21.3 | $\mathbf{A}$ is idempotent $\implies$ 0 and 1 are the only characteristic roots, and $\mathbf{A}$ is positive semidefinite. | |
| 21.4 | $\mathbf{A}$ idempotent with $r$ characteristic roots equal to $1 \implies r(\mathbf{A}) = \text{tr}(\mathbf{A}) = r$ | |
| 21.5 | $\mathbf{A}$ idempotent and $\mathbf{C}$ is orthogonal $\implies \mathbf{C}'\mathbf{A}\mathbf{C}$ is idempotent. | |
| 21.6 | $\mathbf{A}$ idempotent $\iff$ Its associated linear transformation is a projection. | A linear transformation $P$ of $R^n$ into $R^n$ is a *projection* if $P(P(\mathbf{x})) = P(\mathbf{x})$ for all $\mathbf{x}$ in $R^n$. |
| 21.7 | $\mathbf{I}_n - \mathbf{X}(\mathbf{X}'\mathbf{X})^{-1}\mathbf{X}'$ is idempotent. | $\mathbf{X}$ is $n \times m$, $|\mathbf{X}'\mathbf{X}| \neq 0$. |

## Orthogonal matrices

| | | |
|---|---|---|
| 21.8 | $\mathbf{P} = (p_{ij})_{n \times n}$ is *orthogonal* $\iff \mathbf{P}'\mathbf{P} = \mathbf{I}_n$ | Definition of an orthogonal matrix. |
| 21.9 | $\mathbf{P}$ is orthogonal $\iff$ The column vectors of $\mathbf{P}$ are mutually orthogonal unit vectors. | Property of orthogonal matrices. |

| | | |
|---|---|---|
| 21.10 | **P** and **Q** are orthogonal $\Longrightarrow$ **PQ** is orthogonal. | Properties of orthogonal matrices. |
| 21.11 | **P** orthogonal $\Longrightarrow |\mathbf{P}| = \pm 1$, and 1 and $-1$ are the only characteristic roots. | |
| 21.12 | **P** orthogonal $\Leftrightarrow \|\mathbf{Px}\| = \|\mathbf{x}\|$ for all $\mathbf{x} \in R^n$ | Orthogonal transformations preserve lengths of vectors. |

## Permutation matrices

| | | |
|---|---|---|
| 21.13 | $\mathbf{P} = (p_{ij})_{n \times n}$ is a *permutation* matrix if in each row and each column of **P** there is one element equal to 1 and the rest of the elements are 0. | Definition of permutation matrix. |
| 21.14 | **P** permutation matrix $\Longrightarrow$ **P** is nonsingular and orthogonal. | Properties of permutation matrices. |

## Nonnegative matrices

| | | |
|---|---|---|
| 21.15 | $\mathbf{A} = (a_{ij})_{m \times n} \geq \mathbf{0} \iff a_{ij} \geq 0 \quad \text{for all} \quad i,j$ <br> $\mathbf{A} = (a_{ij})_{m \times n} > \mathbf{0} \iff a_{ij} > 0 \quad \text{for all} \quad i,j$ | Definitions of *nonnegative* and *positive* matrices. |
| 21.16 | If $\mathbf{A} = (a_{ij})_{n \times n} \geq 0$, **A** has at least one nonnegative characteristic root. The largest nonnegative characteristic root is called the *Frobenius root* of **A** and it is denoted by $\lambda(\mathbf{A})$. | Definition of the Frobenius root of a matrix. |
| 21.17 | • $\mu$ is a characteristic root of $\mathbf{A} \Longrightarrow |\mu| \leq \lambda(\mathbf{A})$ <br> • $0 \leq \mathbf{A}_1 \leq \mathbf{A}_2 \Longrightarrow \lambda(\mathbf{A}_1) \leq \lambda(\mathbf{A}_2)$ <br> • $\rho > \lambda(\mathbf{A}) \iff (\rho \mathbf{I} - \mathbf{A})^{-1}$ exists and is $\geq 0$ <br> • $\min_{1 \leq j \leq n} \sum_{i=1}^{n} a_{ij} \leq \lambda(\mathbf{A}) \leq \max_{1 \leq j \leq n} \sum_{i=1}^{n} a_{ij}$ | Properties of nonnegative matrices. $\lambda(\mathbf{A})$ is the Frobenius root of $A$. |
| 21.18 | The matrix $\mathbf{A} = (a_{ij})_{n \times n}$ is *decomposable* if its rows and columns can be renumbered such that **A** is transformed to $$\begin{pmatrix} \mathbf{A}_{11} & \mathbf{A}_{12} \\ 0 & \mathbf{A}_{22} \end{pmatrix}$$ where $\mathbf{A}_{11}$ and $\mathbf{A}_{22}$ are square submatrices. | Definition of a decomposable quadratic matrix. A matrix which is not decomposable is called *indecomposable*. |

21.19 $\mathbf{A} = (a_{ij})_{n \times n}$ is decomposable if and only if there exists a permutation matrix $\mathbf{P}$ such that
$$\mathbf{P}^{-1}\mathbf{A}\mathbf{P} = \begin{pmatrix} \mathbf{A}_{11} & \mathbf{A}_{12} \\ 0 & \mathbf{A}_{22} \end{pmatrix}$$
where $\mathbf{A}_{11}$ and $\mathbf{A}_{22}$ are square submatrices.

*A characterization of decomposable matrices.*

21.20 If $\mathbf{A} = (a_{ij})_{n \times n} \geq \mathbf{0}$ is indecomposable,
- the Frobenius root $\lambda(\mathbf{A}) > 0$, and there exists an associated characteristic vector $\mathbf{x} > \mathbf{0}$.
- If $\mathbf{A}\mathbf{x} = \mu \mathbf{x}$ for some $\mu \geq 0$ and $\mathbf{x} \geq \mathbf{0}$, then $\mu = \lambda(\mathbf{A})$.
- $\lambda(\mathbf{A})$ is a simple root of the characteristic polynomial of $\mathbf{A}$.

*Properties of indecomposable matrices.*

## Leontief systems

21.21 If $A = (a_{ij})_{n \times n} \geq 0$, then
$$\mathbf{A}\mathbf{x} + \mathbf{c} = \mathbf{x}$$
is called a *Leontief system*.

*Definition of a Leontief system.* $\mathbf{x}$ and $\mathbf{c}$ are $n \times 1$-matrices.

21.22 If $\sum_{i=1}^{n} a_{ij} < 1$ for $j = 1, \ldots, n$, the Leontief system has a solution $\mathbf{x} \geq \mathbf{0}$.

*Sufficient condition for a Leontief system to have a nonnegative solution.*

21.23 For every $\mathbf{c} \geq \mathbf{0}$, the Leontief system $\mathbf{A}\mathbf{x}+\mathbf{c} = \mathbf{x}$ has a solution $\mathbf{x} \geq \mathbf{0}$ if and only if *one* of the following five equivalent set of conditions are satisfied:
- For some $\mathbf{c} > \mathbf{0}$, $\mathbf{A}\mathbf{x} + \mathbf{c} = \mathbf{x}$ has a solution $\mathbf{x} > \mathbf{0}$.
- The matrix $(\mathbf{I} - \mathbf{A})^{-1}$ exists, is nonnegative and is equal to $\mathbf{I} + \mathbf{A} + \mathbf{A}^2 + \cdots$
- $\mathbf{A}^m \to \mathbf{0}$ as $m \to \infty$.
- All the characteristic roots of $\mathbf{A}$ have moduli $< 1$.
- $\begin{vmatrix} 1-a_{11} & -a_{12} & \cdots & -a_{1k} \\ -a_{21} & 1-a_{12} & \cdots & -a_{2k} \\ \vdots & \vdots & \ddots & \vdots \\ -a_{k1} & -a_{k2} & \cdots & 1-a_{kk} \end{vmatrix} > 0$
  for $k = 1, \ldots, n$.

*Necessary and sufficient conditions for the Leontief system to have a nonnegative solution. The last conditions are the Hawkins-Simon conditions.*

| | | |
|---|---|---|
| 21.24 | $\mathbf{A} = (a_{ij})_{n \times n}$ has a *dominant diagonal* (d.d.) if there exist positive numbers $d_1, \ldots, d_n$ such that $$d_j|a_{jj}| > \sum_{i \neq j} d_i|a_{ij}| \quad \text{for} \quad j = 1, \ldots, n$$ | Definition of a dominant diagonal matrix. |
| 21.25 | If $\mathbf{A}$ has a dominant diagonal, then $|\mathbf{A}| \neq 0$. | A property of dominant diagonal matrices. |
| 21.26 | If $a_{ii} < 1$ for $i = 1, \ldots, n$, while $a_{ij} \geq 0$ for all $i \neq j$, then for each $\mathbf{c} \geq \mathbf{0}$ there exists a solution $\mathbf{x} \geq \mathbf{0}$ of $\mathbf{Ax} + \mathbf{c} = \mathbf{x}$, if and only if $\mathbf{I} - \mathbf{A}$ has a dominant diagonal. | A necessary and sufficient condition for the Leontief system to have a nonnegative solution. |

## References

For the matrix results see Gantmacher (1959). For Leontief systems see Nikaido (1970) and Takayama (1985).

# Chapter 22

# Kronecker products and the vec operator

22.1 $\mathbf{A} \otimes \mathbf{B} = \begin{pmatrix} a_{11}\mathbf{B} & a_{12}\mathbf{B} & \cdots & a_{1n}\mathbf{B} \\ a_{21}\mathbf{B} & a_{22}\mathbf{B} & \cdots & a_{2n}\mathbf{B} \\ \vdots & \vdots & & \vdots \\ a_{m1}\mathbf{B} & a_{m2}\mathbf{B} & \cdots & a_{mn}\mathbf{B} \end{pmatrix}$ | The *Kronecker product* of $\mathbf{A} = (a_{ij})_{m \times n}$ and $\mathbf{B} = (b_{ij})_{p \times q}$. $\mathbf{A} \otimes \mathbf{B}$ is $mp \times nq$.

22.2 $\begin{pmatrix} a_{11} & a_{12} \\ a_{21} & a_{22} \end{pmatrix} \otimes \begin{pmatrix} b_{11} & b_{12} \\ b_{21} & b_{22} \end{pmatrix} = \begin{pmatrix} a_{11}b_{11} & a_{11}b_{12} & a_{12}b_{11} & a_{12}b_{12} \\ a_{11}b_{21} & a_{11}b_{22} & a_{12}b_{21} & a_{12}b_{22} \\ a_{21}b_{11} & a_{21}b_{12} & a_{22}b_{11} & a_{22}b_{12} \\ a_{21}b_{21} & a_{21}b_{22} & a_{22}b_{21} & a_{22}b_{22} \end{pmatrix}$ | A special case of (22.1).

22.3 $\mathbf{A} \otimes \mathbf{B} \otimes \mathbf{C} = (\mathbf{A} \otimes \mathbf{B}) \otimes \mathbf{C} = \mathbf{A} \otimes (\mathbf{B} \otimes \mathbf{C})$ | Valid in general.

22.4 $(\mathbf{A} + \mathbf{B}) \otimes (\mathbf{C} + \mathbf{D}) =$
$\mathbf{A} \otimes \mathbf{C} + \mathbf{A} \otimes \mathbf{D} + \mathbf{B} \otimes \mathbf{C} + \mathbf{B} \otimes \mathbf{D}$ | Valid if $\mathbf{A}+\mathbf{B}$ and $\mathbf{C}+\mathbf{D}$ are defined.

22.5 $(\mathbf{A} \otimes \mathbf{B})(\mathbf{C} \otimes \mathbf{D}) = \mathbf{AC} \otimes \mathbf{BD}$ | Valid if $\mathbf{AC}$ and $\mathbf{BD}$ are defined.

22.6 $(\mathbf{A} \otimes \mathbf{B})' = \mathbf{A}' \otimes \mathbf{B}'$ | Rule for transposing a Kronecker product.

22.7 $(\mathbf{A} \otimes \mathbf{B})^{-1} = \mathbf{A}^{-1} \otimes \mathbf{B}^{-1}$ | Valid if $\mathbf{A}^{-1}$ and $\mathbf{B}^{-1}$ exist.

22.8 $\mathbf{A} \otimes \mathbf{B} \neq \mathbf{B} \otimes \mathbf{A}$ | The Kronecker product is not, in general, commutative.

| | | |
|---|---|---|
| 22.9 | $\operatorname{tr}(\mathbf{A} \otimes \mathbf{B}) = \operatorname{tr}(\mathbf{A}) \operatorname{tr}(\mathbf{B})$ | $\mathbf{A}$ and $\mathbf{B}$ are square matrices, not necessarily of the same order. |
| 22.10 | $\alpha \otimes \mathbf{A} = \alpha \mathbf{A} = \mathbf{A}\alpha = \mathbf{A} \otimes \alpha$ | $\alpha$ is a $1 \times 1$ scalar matrix. |
| 22.11 | If $\lambda_1, \ldots, \lambda_n$ are the characteristic roots of $\mathbf{A}$, and if $\mu_1, \ldots, \mu_p$ are the characteristic roots of $\mathbf{B}$, then $\mathbf{A} \otimes \mathbf{B}$ has the $np$ characteristic roots, $\lambda_i \mu_j$, $i = 1, \ldots, n$, $j = 1, \ldots, p$. | Characteristic roots for $\mathbf{A} \otimes \mathbf{B}$, where $\mathbf{A}$ is $n \times n$ and $\mathbf{B}$ is $p \times p$. |
| 22.12 | If $\mathbf{x}$ is a characteristic vector for $\mathbf{A}$, and $\mathbf{y}$ is a characteristic vector for $\mathbf{B}$, then $\mathbf{x} \otimes \mathbf{y}$ is a characteristic vector for $\mathbf{A} \otimes \mathbf{B}$. | Note that a characteristic vector for $\mathbf{A} \otimes \mathbf{B}$ is not necessarily the Kronecker product of a characteristic vector of $\mathbf{A}$ and a characteristic vector of $\mathbf{B}$. |
| 22.13 | If $\mathbf{A}$ and $\mathbf{B}$ are positive (semi-)definite, then $\mathbf{A} \otimes \mathbf{B}$ is positive (semi-)definite. | Follows from (22.11). |
| 22.14 | $\lvert \mathbf{A} \otimes \mathbf{B} \rvert = \lvert \mathbf{A} \rvert^p \cdot \lvert \mathbf{B} \rvert^n$ | $\mathbf{A}$ is $n \times n$, $\mathbf{B}$ is $p \times p$. |
| 22.15 | $r(\mathbf{A} \otimes \mathbf{B}) = r(\mathbf{A})\, r(\mathbf{B})$ | The rank of a Kronecker product. |
| 22.16 | If $\mathbf{A} = [\mathbf{a}_1, \mathbf{a}_2, \ldots, \mathbf{a}_n]_{m \times n}$, then $$\operatorname{vec}(\mathbf{A}) = \begin{pmatrix} \mathbf{a}_1 \\ \mathbf{a}_2 \\ \vdots \\ \mathbf{a}_n \end{pmatrix}_{(mn \times 1)}$$ | Definition of $\operatorname{vec}(\mathbf{A})$ for a matrix $\mathbf{A}$. |
| 22.17 | $\operatorname{vec} \begin{pmatrix} a_{11} & a_{12} \\ a_{21} & a_{22} \end{pmatrix} = \begin{pmatrix} a_{11} \\ a_{21} \\ a_{12} \\ a_{22} \end{pmatrix}$ | A special case of (22.16). |
| 22.18 | $\operatorname{vec}(\mathbf{A} + \mathbf{B}) = \operatorname{vec}(\mathbf{A}) + \operatorname{vec}(\mathbf{B})$ | Valid if $\mathbf{A} + \mathbf{B}$ is defined. |
| 22.19 | $\operatorname{vec}(\mathbf{ABC}) = [\mathbf{C}' \otimes \mathbf{A}]\, \operatorname{vec}(\mathbf{B})$ | Valid if the product $\mathbf{ABC}$ is defined. |

22.20 $\operatorname{tr}(\mathbf{AB}) = (\operatorname{vec}(\mathbf{A}'))' \operatorname{vec}(\mathbf{B}) = (\operatorname{vec}(\mathbf{B}'))' \operatorname{vec}(\mathbf{A})$ | Valid if the operations are defined.

## References

See Magnus and Neudecker (1988) and Dhrymes (1978).

# Chapter 23

# Differentiation of vectors and matrices

23.1  If $y = f(x_1, \ldots, x_n) = f(\mathbf{x})$, then
$$\frac{\partial y}{\partial \mathbf{x}} = \left( \frac{\partial y}{\partial x_1}, \ldots, \frac{\partial y}{\partial x_n} \right)$$

The gradient of $y = f(\mathbf{x})$. (A row vector. The derivative of a scalar function of a vector variable.)

23.2  $\begin{aligned} y_1 &= f_1(x_1, \ldots, x_n) \\ &\cdots \\ y_m &= f_m(x_1, \ldots, x_n) \end{aligned} \iff \mathbf{y} = \mathbf{f}(\mathbf{x})$

A transformation $\mathbf{f}$ from $R^n$ to $R^m$. We let $\mathbf{x}$ and $\mathbf{y}$ be column vectors.

23.3  $\dfrac{\partial \mathbf{y}}{\partial \mathbf{x}} = \begin{pmatrix} \dfrac{\partial y_1(\mathbf{x})}{\partial x_1} & \cdots & \dfrac{\partial y_1(\mathbf{x})}{\partial x_n} \\ \vdots & & \vdots \\ \dfrac{\partial y_m(\mathbf{x})}{\partial x_1} & \cdots & \dfrac{\partial y_m(\mathbf{x})}{\partial x_n} \end{pmatrix}$

The *Jacobian matrix* of the transformation in (23.2). (The derivative of a vector function of a vector variable.)

23.4  $\dfrac{\partial^2 \mathbf{y}}{\partial \mathbf{x} \partial \mathbf{x}'} = \dfrac{\partial}{\partial \mathbf{x}} \operatorname{vec}\left[ \left( \dfrac{\partial \mathbf{y}}{\partial \mathbf{x}} \right)' \right]$

A definition. For the vec operator see (22.16).

23.5  $\dfrac{\partial \mathbf{A}(\mathbf{r})}{\partial \mathbf{r}} = \dfrac{\partial}{\partial \mathbf{r}} \operatorname{vec}(\mathbf{A}(\mathbf{r}))$

A general definition of the derivative of a matrix with respect to a vector.

23.6  $\dfrac{\partial^2 y}{\partial \mathbf{x} \partial \mathbf{x}'} = \begin{pmatrix} \dfrac{\partial^2 y}{\partial x_1^2} & \cdots & \dfrac{\partial^2 y}{\partial x_n \partial x_1} \\ \vdots & \ddots & \vdots \\ \dfrac{\partial^2 y}{\partial x_1 \partial x_n} & \cdots & \dfrac{\partial^2 y}{\partial x_n^2} \end{pmatrix}$

A special case of (23.4). ($\partial^2 y / \partial \mathbf{x} \partial \mathbf{x}'$ is the Hessian matrix defined in (12.19).)

23.7 $\quad\dfrac{\partial}{\partial \mathbf{x}}(\mathbf{a}' \cdot \mathbf{x}) = \mathbf{a}'$  $\qquad$ $\mathbf{a}$ and $\mathbf{x}$ are $n \times 1$-vectors.

23.8 $\quad\begin{aligned}\dfrac{\partial}{\partial \mathbf{x}}(\mathbf{x}'\mathbf{A}\mathbf{x}) &= \mathbf{x}'(\mathbf{A}+\mathbf{A}') \\ \dfrac{\partial^2}{\partial \mathbf{x}\partial \mathbf{x}'}(\mathbf{x}'\mathbf{A}\mathbf{x}) &= \mathbf{A}+\mathbf{A}'\end{aligned}$  $\qquad$ Differentiation of a quadratic form. $\mathbf{A}$ is $n \times n$, $\mathbf{x}$ is $n \times 1$.

23.9 $\quad\dfrac{\partial}{\partial \mathbf{x}}(\mathbf{A}\mathbf{x}) = \mathbf{A}$  $\qquad$ $\mathbf{A}$ is $m \times n$, $\mathbf{x}$ is $n \times 1$.

23.10 If $\mathbf{y} = \mathbf{A}(\mathbf{r})\mathbf{x}(\mathbf{r})$, then
$$\dfrac{\partial \mathbf{y}}{\partial \mathbf{r}} = (\mathbf{x}' \otimes I_m)\dfrac{\partial \mathbf{A}}{\partial \mathbf{r}} + \mathbf{A}\dfrac{\partial \mathbf{x}}{\partial \mathbf{r}}$$
$\mathbf{A}(\mathbf{r})$ is $m \times n$, $\mathbf{x}(\mathbf{r})$ is $n \times 1$ and $\mathbf{r}$ is $k \times 1$.

23.11 If $y = f(\mathbf{A})$, then
$$\dfrac{\partial y}{\partial \mathbf{A}} = \begin{pmatrix} \dfrac{\partial y}{\partial a_{11}} & \cdots & \dfrac{\partial y}{\partial a_{1n}} \\ \vdots & & \vdots \\ \dfrac{\partial y}{\partial a_{m1}} & \cdots & \dfrac{\partial y}{\partial a_{mn}} \end{pmatrix}$$
The derivative of a scalar function of a matrix $A = (a_{ij})_{m \times n}$.

23.12 $\quad\dfrac{\partial |\mathbf{A}|}{\partial \mathbf{A}} = (A_{ij})$  $\qquad$ $\mathbf{A}$ is $n \times n$. $(A_{ij})$ is the matrix of cofactors of $\mathbf{A}$. (See (19.16).)

23.13 $\quad\dfrac{\partial \operatorname{tr}(\mathbf{A})}{\partial \mathbf{A}} = \mathbf{I}_n$  $\qquad$ $\mathbf{A}$ is $n \times n$.

23.14 $\quad\dfrac{\partial a^{ij}}{\partial a_{hk}} = -a^{ih}a^{kj}; \qquad i,j,h,k = 1,\ldots,n$  $\qquad$ $a^{ij}$ is the $ij$th element of $\mathbf{A}^{-1}$.

## References

See Dhrymes (1978) and Magnus and Neudecker (1988).

# Chapter 24

# Comparative statics. Value functions

| | | |
|---|---|---|
| 24.1 | $E_1(\mathbf{p},\mathbf{a}) = S_1(\mathbf{p},\mathbf{a}) - D_1(\mathbf{p},\mathbf{a})$ <br> $E_2(\mathbf{p},\mathbf{a}) = S_2(\mathbf{p},\mathbf{a}) - D_2(\mathbf{p},\mathbf{a})$ <br> $\dots\dots\dots\dots\dots\dots\dots\dots\dots\dots$ <br> $E_n(\mathbf{p},\mathbf{a}) = S_n(\mathbf{p},\mathbf{a}) - D_n(\mathbf{p},\mathbf{a})$ | $S_i(\mathbf{p},\mathbf{a})$ is supply and $D_i(\mathbf{p},\mathbf{a})$ is demand for good $i$. $E_i(\mathbf{p},\mathbf{a})$ is excess supply. $\mathbf{p} = (p_1,\dots,p_n)$ is the price vector, $\mathbf{a} = (a_1,\dots,a_k)$ is a vector of exogenous variables. |
| 24.2 | $E_1(\mathbf{p},\mathbf{a}) = 0, \ E_2(\mathbf{p},\mathbf{a}) = 0, \ \dots, \ E_n(\mathbf{p},\mathbf{a}) = 0$ | Conditions for equilibrium. |
| 24.3 | $E_1(p_1,p_2,a_1,\dots,a_k) = 0$ <br> $E_2(p_1,p_2,a_1,\dots,a_k) = 0$ | Equilibrium conditions for the two good case. |
| 24.4 | $\dfrac{\partial p_1}{\partial a_j} = \dfrac{\dfrac{\partial E_1}{\partial p_2}\dfrac{\partial E_2}{\partial a_j} - \dfrac{\partial E_2}{\partial p_2}\dfrac{\partial E_1}{\partial a_j}}{\dfrac{\partial E_1}{\partial p_1}\dfrac{\partial E_2}{\partial p_2} - \dfrac{\partial E_1}{\partial p_2}\dfrac{\partial E_2}{\partial p_1}}$ <br><br> $\dfrac{\partial p_2}{\partial a_j} = \dfrac{\dfrac{\partial E_2}{\partial p_1}\dfrac{\partial E_1}{\partial a_j} - \dfrac{\partial E_1}{\partial p_1}\dfrac{\partial E_2}{\partial a_j}}{\dfrac{\partial E_1}{\partial p_1}\dfrac{\partial E_2}{\partial p_2} - \dfrac{\partial E_1}{\partial p_2}\dfrac{\partial E_2}{\partial p_1}}$ | Comparative statics results for the two good case, $j = 1,\dots,k$. |
| 24.5 | $\begin{pmatrix} \dfrac{\partial p_1}{\partial a_j} \\ \vdots \\ \dfrac{\partial p_n}{\partial a_j} \end{pmatrix} = - \begin{pmatrix} \dfrac{\partial E_1}{\partial p_1} & \cdots & \dfrac{\partial E_1}{\partial p_n} \\ \vdots & \ddots & \vdots \\ \dfrac{\partial E_n}{\partial p_1} & \cdots & \dfrac{\partial E_n}{\partial p_n} \end{pmatrix}^{-1} \begin{pmatrix} \dfrac{\partial E_1}{\partial a_j} \\ \vdots \\ \dfrac{\partial E_n}{\partial a_j} \end{pmatrix}$ | Comparative statics results for the $n$ good case, $j = 1,\dots,k$. See (19.16) for the general formula for the inverse of a square matrix. |

24.6 $\quad V(\mathbf{a}) = \max_{\mathbf{x} \in X} f(\mathbf{x}, \mathbf{a})$

*The value function* for a maximization problem. $X$ subset of $R^n$, $\mathbf{a} \in A$, $A$ a subset of $R^k$.

24.7 If $f(\mathbf{x}, \mathbf{a})$ is continuous on $X \times A$ and $X$ is compact, then $V(\mathbf{a})$ defined in (24.6) is continuous on $A$. If the problem in (24.6) has a unique solution $\mathbf{x} = \mathbf{x}(\mathbf{a})$ for each $\mathbf{a} \in A$, then $\mathbf{x}(\mathbf{a})$ is a continuous function of $\mathbf{a}$.

Continuity of the value function and its maximizer.

24.8 Suppose that problem (24.6) has a unique solution $\mathbf{x}^0(\mathbf{a}^0)$ at $\mathbf{a} = \mathbf{a}^0$, and that $\partial f/\partial a_i$, $i = 1, \ldots, k$, exist and are continuous in a neigborhood of $(\mathbf{x}^0, \mathbf{a}^0)$. Then for $i = 1, \ldots, k$,
$$\frac{\partial V(\mathbf{a}^0)}{\partial a_i} = \frac{\partial f(\mathbf{x}^0, \mathbf{a}^0)}{\partial a_i}$$

An *envelope theorem*.

24.9 $\quad \max_{\mathbf{x}} f(\mathbf{x}, \mathbf{a})$ when $g_j(\mathbf{x}, \mathbf{a}) = 0$, $j = 1, \ldots, m$

A Lagrange problem with parameters, $\mathbf{a} = (a_1, \ldots, a_k)$.

24.10 $\quad V(\mathbf{a}) = \sup \{f(\mathbf{x}, \mathbf{a}) : g_j(\mathbf{x}, \mathbf{a}) = 0, \, j = 1, \ldots, m\}$

The *value function* for problem (24.9).

24.11 Consider problem (24.9) and suppose that
- for $\mathbf{a} = \mathbf{a}^0$ the problem has a unique solution $\mathbf{x}^0 = \mathbf{x}(\mathbf{a}^0)$;
- there exists a ball $B(\mathbf{a}^0; \alpha)$ and a constant $K$ such that for every $\mathbf{a} \in B(\mathbf{a}^0; \alpha)$ there exist solutions $\mathbf{x}'$ of (24.9) such that $\mathbf{x}' \in B(\mathbf{x}^0; K)$;
- $f$ and $g_1, \ldots, g_m$ are $C^1$-functions in some ball around $(\mathbf{x}(\mathbf{a}^0), \mathbf{a}^0)$;
- the matrix $(\partial g_j(\mathbf{x}^0)/\partial x_i)_{m \times n}$ has rank $m$.

Then $V(\mathbf{a})$ has partial derivatives at $\mathbf{a}^0$, and
$$\frac{\partial V(\mathbf{a}^0)}{\partial a_i} = \frac{\partial L(\mathbf{x}(\mathbf{a}^0), \mathbf{a}^0)}{\partial a_i}, \quad i = 1, \ldots, k.$$

An envelope theorem for problem (24.9). $L = f - \sum \lambda_j g_j$ is the Lagrange function. (See Chapter 13.)

24.12 $\quad \max_{\mathbf{x}} f(\mathbf{x}, \mathbf{a})$ when $g_j(\mathbf{x}, \mathbf{a}) \leq 0$, $j = 1, \ldots, m$

A nonlinear programming problem with parameters, $\mathbf{a} = (a_1, \ldots, a_k)$.

24.13 $\quad V(\mathbf{a}) = \sup \{f(\mathbf{x}, \mathbf{a}) : g_j(\mathbf{x}, \mathbf{a}) \leq 0, \, j = 1, \ldots, m\}$

The *value function* for problem (24.12).

24.14

Consider problem (24.12) and suppose that
- for $\mathbf{a} = \mathbf{a}^0$ the problem has a unique solution $\mathbf{x}^0 = \mathbf{x}(\mathbf{a}^0)$;
- there exists a ball $B(\mathbf{a}^0; \alpha)$ and a constant $K$ such that for every $\mathbf{a} \in B(\mathbf{a}^0; \alpha)$ there exist solutions $\mathbf{x}'$ of (24.12) such that $\mathbf{x}' \in B(\mathbf{x}^0; K)$;
- $f$ and $g_1, \ldots, g_m$ are $C^1$-functions in a certain ball around $(\mathbf{x}(\mathbf{a}^0), \mathbf{a}^0)$;
- the gradients at $\mathbf{x}^0$ of those $g_j$-functions whose constraints are active at $\mathbf{x}^0$ are linearly independent.

Then $V(\mathbf{a})$ has partial derivatives at $\mathbf{a}^0$, and
$$\frac{\partial V(\mathbf{a}^0)}{\partial a_i} = \frac{\partial \mathcal{L}(\mathbf{x}(\mathbf{a}^0), \mathbf{a}^0, \lambda)}{\partial a_i}, \quad i = 1, \ldots, k.$$

An *envelope theorem* for problem (24.12). $\mathcal{L} = f - \sum \lambda_j g_j$ is the Lagrange function, and constraint $j$ is active at $(\mathbf{x}^0, \mathbf{a}^0)$ if $g_j(\mathbf{x}^0, \mathbf{a}^0) = 0$. (See Chapter 14.)

## References

On comparative statics see Varian (1989) or Silberberg (1978). For properties of value functions see Dixit (1990).

## Chapter 25

# Properties of cost and profit functions

25.1 $\quad C(\mathbf{w}, y) = \min_{\mathbf{x}} \sum_{i=1}^{n} w_i x_i \quad \text{when} \quad f(\mathbf{x}) = y$

*Cost minimization.* One output. $f$ is the production function, $\mathbf{w} = (w_1, \ldots, w_n)$ are factor prices, $y$ is output and $\mathbf{x} = (x_1, \ldots, x_n)$ are factor inputs. $C(\mathbf{w}, y)$ is the cost function.

25.2 $\quad C(\mathbf{w}, y) = \begin{cases} \text{The minimum cost of producing} \\ y \text{ units of a commodity when} \\ \text{factor prices are } \mathbf{w} = (w_1, \ldots, w_n) \end{cases}$

The *cost function*.

25.3 $\quad \begin{cases} C(\mathbf{w}, y) \text{ is nondecreasing in each } w_i. \\ C(\mathbf{w}, y) \text{ is homogeneous of degree 1 in } \mathbf{w}. \\ C(\mathbf{w}, y) \text{ is concave in } \mathbf{w}. \\ C(\mathbf{w}, y) \text{ is continuous in } \mathbf{w} \text{ for } \mathbf{w} > \mathbf{0}. \end{cases}$

Properties of the cost function.

25.4 $\quad x_i(\mathbf{w}, y) = \begin{cases} \text{The cost minimizing choice of the} \\ i\text{th input factor as a function of} \\ \text{the factor prices } \mathbf{w} \text{ and the} \\ \text{production level } y. \end{cases}$

*Conditional factor demand functions.* $\mathbf{x}(\mathbf{w}, y)$ is the vector $\mathbf{x}$ which solves the problem in (25.1).

25.5 $\quad \begin{cases} x_i(\mathbf{w}, y) \text{ is nonincreasing in } w_i. \\ x_i(\mathbf{w}, y) \text{ is homogeneous of degree 0 in } \mathbf{w}. \end{cases}$

Properties of the conditional factor demand function.

25.6 $\quad \dfrac{\partial C(\mathbf{w}, y)}{\partial w_i} = x_i(\mathbf{w}, y), \quad i = 1, \ldots, n$

*Shephard's lemma.*

| | | |
|---|---|---|
| 25.7 | $\left(\dfrac{\partial^2 C(\mathbf{w},y)}{\partial w_i \partial w_j}\right)_{(n\times n)} = \left(\dfrac{\partial x_i(\mathbf{w},y)}{\partial w_j}\right)_{(n\times n)}$ is symmetric and negative semidefinite. | Properties of the *substitution matrix*. |
| 25.8 | $\pi(p,\mathbf{w}) = \max_{\mathbf{x}} \left( pf(\mathbf{x}) - \sum_{i=1}^{n} w_i x_i \right)$ | The profit maximizing problem of the firm. $p$ is the price of output. $\pi(p,\mathbf{w})$ is the *profit function*. |
| 25.9 | $\pi(p,\mathbf{w}) = \begin{cases} \text{The maximum profit of the firm} \\ \text{as a function of the factor prices} \\ \mathbf{w} = (w_1,\ldots,w_n) \text{ and the price} \\ \text{of output } p. \end{cases}$ | The profit function. |
| 25.10 | $\pi(p,\mathbf{w}) \equiv \max_{y}(py - C(\mathbf{w},y))$ | The profit function in terms of costs and revenue. |
| 25.11 | $\begin{cases} \pi(p,\mathbf{w}) \text{ is increasing in } p. \\ \pi(p,\mathbf{w}) \text{ is homogeneous of degree 1 in } (p,\mathbf{w}). \\ \pi(p,\mathbf{w}) \text{ is convex in } (p,\mathbf{w}). \\ \pi(p,\mathbf{w}) \text{ is continuous in } (p,\mathbf{w}) \text{ for } \mathbf{w}>\mathbf{0}, p>0. \end{cases}$ | Properties of the profit function. |
| 25.12 | $x_i(p,\mathbf{w}) = \begin{cases} \text{The profit maximizing choice of} \\ \text{the }i\text{th input factor as a function} \\ \text{of the price of output }p \text{ and the} \\ \text{factor prices } \mathbf{w} \end{cases}$ | The *factor demand functions*. $\mathbf{x}(\mathbf{w},p)$ is the vector $\mathbf{x}$ which solves the problem in (25.8). |
| 25.13 | $\begin{cases} x_i(p,\mathbf{w}) \text{ is nonincreasing in } w_i. \\ x_i(p,\mathbf{w}) \text{ is homogeneous of degree 0 in } (p,\mathbf{w}). \\ \text{The cross-price effects are symmetric:} \\ \dfrac{\partial x_i(p,\mathbf{w})}{\partial w_j} = \dfrac{\partial x_j(p,\mathbf{w})}{\partial w_i}, \quad i,j=1,\ldots,n \end{cases}$ | Properties of the factor demand functions. |
| 25.14 | $y(p,\mathbf{w}) = \begin{cases} \text{The profit maximizing output} \\ \text{as a function of the price of output} \\ p \text{ and the factor prices } \mathbf{w}. \end{cases}$ | The *supply function* $f(\mathbf{x}(p,\mathbf{w}))$ is the $y$ which solves the problem in (25.10). |

25.15 $\begin{cases} y(p,\mathbf{w}) \text{ is nondecreasing in } p. \\ y(p,\mathbf{w}) \text{ is homogeneous of degree } 0 \text{ in } (p,\mathbf{w}). \end{cases}$ — Properties of the supply function.

25.16 $\dfrac{\partial \pi(p,\mathbf{w})}{\partial p} = y(p,\mathbf{w})$

$\dfrac{\partial \pi(p,\mathbf{w})}{\partial w_i} = -x_i(p,\mathbf{w}), \quad i = 1,\ldots,n$ — Hotelling's lemma.

25.17 $\dfrac{\partial x_j(p,\mathbf{w})}{\partial w_k} = \dfrac{\partial x_j(\mathbf{w},y)}{\partial w_k} + \dfrac{\dfrac{\partial x_j(p,\mathbf{w})}{\partial p}\dfrac{\partial y(p,\mathbf{w})}{\partial w_k}}{\dfrac{\partial y(p,\mathbf{w})}{\partial p}}$ — Puu's equation, $j,k = 1,\ldots,n$. The equation shows the substitution and the scale effects of an increase in the factor price.

## Special functional forms and their properties

### The Cobb-Douglas production function

25.18 $y = A x_1^{a_1} x_2^{a_2} \ldots x_n^{a_n}$ — The Cobb-Douglas production function defined for $x_i > 0$, $i = 1,\ldots,n$. $a_1, \ldots, a_n$, and $A$ are positive constants.

25.19 The Cobb-Douglas function in (25.18) is:
(a) homogeneous of degree $a_1 + \cdots + a_n$,
(b) quasi-concave for all $a_1, \ldots, a_n$,
(c) concave for $a_1 + \cdots + a_n \leq 1$, and
(d) strictly concave for $a_1 + \cdots + a_n < 1$.
— Properties of the Cobb-Douglas function. ($a_1, \ldots, a_n$, and $A$ are positive constants.)

25.20 $x_k = A^{-1/s}\left(\dfrac{a_k}{w_k}\right)\left(\dfrac{w_1}{a_1}\right)^{a_1/s} \cdots \left(\dfrac{w_n}{a_n}\right)^{a_n/s} y^{1/s}$ — Conditional factor demand functions with $s = a_1 + \cdots + a_n$.

25.21 $C(\mathbf{w},y) = s A^{-1/s}\left(\dfrac{w_1}{a_1}\right)^{a_1/s} \cdots \left(\dfrac{w_n}{a_n}\right)^{a_n/s} y^{1/s}$ — The cost function with $s = a_1 + \cdots + a_n$.

25.22 $\dfrac{w_k x_k}{C(\mathbf{w},y)} = \dfrac{a_k}{a_1 + \cdots + a_n}$ — Factor shares in total costs.

25.23 $\pi = (1-s)(pA)^{1/(1-s)} \prod_{i=1}^{n} \left(\frac{w_i}{a_i}\right)^{-a_i/(1-s)}$     The profit function with $s = a_1 + \cdots + a_n < 1$.

## The CES production function

25.24 $y = \left((a_1 x_1)^e + (a_2 x_2)^e + \cdots + (a_n x_n)^e\right)^{1/e}$     The CES production function defined for $x_i > 0$, $i = 1, \ldots, n$. $a_1, \ldots, a_n$ are positive, $e \neq 0$, $e < 1$.

25.25 The CES-function in (25.24) is:
(a) homogeneous of degree 1,
(b) concave for $e \leq 1$,
(c) convex for $e \geq 1$
    Properties of the CES-function.

25.26 $x_k = w_k^{r-1} a_k^{-r} \left(\left(\frac{w_1}{a_1}\right)^r + \cdots + \left(\frac{w_n}{a_n}\right)^r\right)^{-1/e} y$     Conditional factor demand functions with $r = e/(e-1)$.

25.27 $C(\mathbf{w}, y) = \left(\left(\frac{w_1}{a_1}\right)^r + \cdots + \left(\frac{w_n}{a_n}\right)^r\right)^{1/r} y$     The cost function with $r = e/(e-1)$.

25.28 $\dfrac{w_k x_k}{C(\mathbf{w}, y)} = \dfrac{\left(\dfrac{w_k}{a_k}\right)^r}{\left(\dfrac{w_1}{a_1}\right)^r + \cdots + \left(\dfrac{w_n}{a_n}\right)^r}$     Factor shares in total costs.

## Law of the minimum

25.29 $y = \min(a_1 + b_1 x_1, \ldots, a_n + b_n x_n)$     Law of the minimum. When $a_1 = \ldots = a_n = 0$, this is the *Leontief* or *fixed coefficient function*.

25.30 $x_k = \dfrac{y - a_k}{b_k}, \quad k = 1, \ldots, n$     Conditional factor demand functions.

25.31 $C(\mathbf{w}, y) = \left(\dfrac{y - a_1}{b_1}\right) w_1 + \cdots + \left(\dfrac{y - a_n}{b_n}\right) w_n$     The cost function.

## The Translog cost function

25.32
$$\ln C(\mathbf{w}, y) = a_0 + c_1 \ln y + \sum_{i=1}^{n} a_i \ln w_i \\ + \frac{1}{2} \sum_{i,j=1}^{n} a_{ij} \ln w_i \ln w_j + \sum_{i=1}^{n} b_i \ln w_i \ln y$$

The translog cost function, $a_{ij} = a_{ji}$ for all $i$ and $j$.

25.33
$$\sum_{i=1}^{n} a_i = 1, \quad \sum_{i=1}^{n} b_i = 0 \\ \sum_{j=1}^{n} a_{ij} = 0, \quad i = 1, \ldots, n \\ \sum_{i=1}^{n} a_{ij} = 0, \quad j = 1, \ldots, n$$

Conditions for the translog function to be homogeneous of degree one in $\mathbf{w}$.

25.34
$$\frac{w_k x_k}{C(\mathbf{w}, y)} = a_k + \sum_{j=1}^{n} a_{kj} \ln w_j + b_i \ln y$$

Factor shares in total costs.

## References

The basic reference is Varian (1989). For a detailed discussion of existence and differentiability assumptions, see Fuss and McFadden (1978). For a discussion on the Puu equation (25.17), see Johansen (1972).

# Chapter 26

## Consumer theory

26.1    $\max\limits_{\mathbf{x}} u(\mathbf{x})$ subject to $\sum_{i=1}^{n} p_i x_i = m$   |   *Utility maximization.* $\mathbf{x} = (x_1, \cdots, x_n)$ is a vector of commodities, $\mathbf{p} = (p_1, \ldots, p_n)$ is the price vector, $m$ is income, and $u$ is the utility function.

26.2    $v(\mathbf{p}, m) = \begin{cases} \text{The maximum utility obtained as} \\ \text{a function of the price vector } \mathbf{p} \\ \text{and income } m. \end{cases}$   |   The *indirect utility function.* $v(\mathbf{p}, m) = \max\limits_{\mathbf{x}} \{u(\mathbf{x}) : \mathbf{p} \cdot \mathbf{x} = m\}$

26.3    $\begin{cases} v(\mathbf{p}, m) \text{ is nonincreasing in } \mathbf{p} \\ v(\mathbf{p}, m) \text{ is nondecreasing in } m \\ v(\mathbf{p}, m) \text{ is homogeneous of degree 0 in } (\mathbf{p}, m) \\ v(\mathbf{p}, m) \text{ is quasi-convex in } \mathbf{p} \\ v(\mathbf{p}, m) \text{ is continuous in } (\mathbf{p}, m), \mathbf{p} > 0, m > 0 \end{cases}$   |   Properties of the indirect utility function.

26.4    $x_i(\mathbf{p}, m) = \begin{cases} \text{The optimal choice of the } i\text{th} \\ \text{commodity as a function of the} \\ \text{price vector } \mathbf{p} \text{ and the income } m. \end{cases}$   |   The *consumer demand functions*, or *Marshallian demand functions*. $\mathbf{x}(\mathbf{p}, m)$ is the vector $\mathbf{x}$ which solves problem (26.1).

26.5    $\mathbf{x}(t\mathbf{p}, tm) = \mathbf{x}(\mathbf{p}, m)$,    $t$ is a positive scalar   |   The consumer demand function is homogeneous of degree 0.

26.6    $x_i(\mathbf{p}, m) = -\dfrac{\dfrac{\partial v(\mathbf{p}, m)}{\partial p_i}}{\dfrac{\partial v(\mathbf{p}, m)}{\partial m}}$,    $i = 1, \ldots, n$   |   Roy's identity.

| | | |
|---|---|---|
| 26.7 | $e(\mathbf{p}, u) = \min_{\mathbf{x}} \{\mathbf{p} \cdot \mathbf{x} : u(\mathbf{x}) \geq u\} =$ the minimum expenditure at prices $\mathbf{p}$ for obtaining at least the utility level $u$. | The *expenditure function*. |
| 26.8 | $\begin{cases} e(\mathbf{p}, u) \text{ is nondecreasing in } \mathbf{p}. \\ e(\mathbf{p}, u) \text{ is homogeneous of degree 1 in } \mathbf{p}. \\ e(\mathbf{p}, u) \text{ is concave in } \mathbf{p}. \\ e(\mathbf{p}, u) \text{ is continuous in } \mathbf{p} \text{ for } \mathbf{p} > \mathbf{0}. \end{cases}$ | Properties of the expenditure function. |
| 26.9 | $\mathbf{h}(\mathbf{p}, u) = \begin{cases} \text{The expenditure-minimizing bundle} \\ \text{necessary to achieve utility level } u \\ \text{at prices } \mathbf{p}. \end{cases}$ | The *Hicksian (or compensated) demand function*. $\mathbf{h}(\mathbf{p}, u)$ is the vector $\mathbf{x}$ which solves the problem $\min_{\mathbf{x}}\{\mathbf{p} \cdot \mathbf{x} : u(\mathbf{x}) \geq u\}$. |
| 26.10 | $\dfrac{\partial e(\mathbf{p}, u)}{\partial p_i} = h_i(\mathbf{p}, u), \quad \text{for} \quad i = 1, \ldots, n$ | Assume that $e$ is differentiable for $\mathbf{p} > \mathbf{0}$. |
| 26.11 | $\dfrac{\partial h_i(\mathbf{p}, u)}{\partial p_j} = \dfrac{\partial h_j(\mathbf{p}, u)}{\partial p_i}, \quad i, j = 1, \ldots, n$ | Symmetry of the Hicksian cross partials. (The *Marshallian cross partials* need not be symmetric.) |
| 26.12 | The matrix of substitution terms $\left(\dfrac{\partial h_i(\mathbf{p}, u)}{\partial p_j}\right)$ is negative semidefinite. | Follows from (26.10) and the concavity of the expenditure function. |
| 26.13 | $e(\mathbf{p}, v(\mathbf{p}, m)) = m : \begin{cases} \text{The minimal expenditure} \\ \text{to reach utility } v(\mathbf{p}, m) \text{ is } m. \end{cases}$ | Identities valid in non-perverse cases. |
| 26.14 | $v(\mathbf{p}, e(\mathbf{p}, u)) = u : \begin{cases} \text{The maximal utility from} \\ \text{income } e(\mathbf{p}, u) \text{ is } u. \end{cases}$ | |
| 26.15 | $x_i(\mathbf{p}, m) = h_i(\mathbf{p}, v(\mathbf{p}, m)) : \begin{cases} \text{Marshallian demand} \\ \text{at income } m \text{ is} \\ \text{Hicksian demand at} \\ \text{utility } v(\mathbf{p}, m) \end{cases}$ | |

26.16  $h_i(\mathbf{p}, u) = x_i(\mathbf{p}, e(\mathbf{p}, u)):$ $\begin{cases} \text{Hicksian demand at} \\ \text{utility } u \text{ equals} \\ \text{Marshallian demand} \\ \text{at income } e(\mathbf{p}, u). \end{cases}$   Identity valid in nonperverse cases.

26.17  $\dfrac{\partial x_i(\mathbf{p}, m)}{\partial p_j} = \dfrac{\partial h_i(\mathbf{p}, u)}{\partial p_j} - x_j(\mathbf{p}, m)\dfrac{\partial x_i(\mathbf{p}, m)}{\partial m}$   The *Slutsky equation*.

26.18  The following $\frac{1}{2}n(n+1) + 1$ restrictions on the partial derivatives of the demand functions are linearly independent:

(a) $\sum_{i=1}^{n} p_i \dfrac{\partial x_i(\mathbf{p}, m)}{\partial m} = 1$

(b) $\sum_{j=1}^{n} p_j \dfrac{\partial x_i}{\partial p_j} + m\dfrac{\partial x_i}{\partial m} = 0, \quad i = 1, \ldots, n$

(c) $\dfrac{\partial x_i}{\partial p_j} + x_j\dfrac{\partial x_i}{\partial m} = \dfrac{\partial x_j}{\partial p_i} + x_i\dfrac{\partial x_j}{\partial m}$, for $i = 1, \ldots, n-1, \ j = i+1, \ldots, n$

(a) is the budget constraint differentiated with respect to $m$.
(b) is the Euler equation (for homogeneous functions) applied to the consumer demand function.
(c) is a consequence of the Slutsky equation.

26.19  $EV = e(\mathbf{p}^0, v(\mathbf{p}^1, m^1)) - e(\mathbf{p}^0, v(\mathbf{p}^0, m^0))$

$EV$ is the difference between the amount of money needed at the old (period 0) prices to reach the new (period 1) utility level, and the amount of money at the old prices needed to reach the old utility level.

*Equivalent variation.* $\mathbf{p}^0$, $m^0$, and $\mathbf{p}^1$, $m^1$ are prices and income in period 0 and 1, respectively. $(e(\mathbf{p}^0, v(\mathbf{p}^0, m^0)) = m^0)$

26.20  $CV = e(\mathbf{p}^1, v(\mathbf{p}^1, m^1)) - e(\mathbf{p}^1, v(\mathbf{p}^0, m^0))$

$CV$ is the difference between the amount of money needed at the new (period 1) prices to reach the new utility level, and the amount of money at the new prices needed to reach the old (period 0) utility level.

*Compensating variation.* $\mathbf{p}^0$, $m^0$, and $\mathbf{p}^1$, $m^1$ are prices and income in period 0 and 1, respectively. $(e(\mathbf{p}^1, v(\mathbf{p}^1, m^1)) = m^1)$

## Special functional forms and their properties

### Almost Ideal Demand System (AIDS)

26.21
$$\ln(e(\mathbf{p}, u)) = a(\mathbf{p}) + ub(\mathbf{p})$$
$$a(\mathbf{p}) = \alpha_0 + \sum_{i=1}^{n} \alpha_i \ln p_i + \frac{1}{2} \sum_{i,j=1}^{n} \gamma_{ij}^* \ln p_i \ln p_j$$
$$b(\mathbf{p}) = \beta_0 \prod_{i=1}^{n} p_i^{\beta_i}$$

Almost ideal demand system, defined by the logarithm of the expenditure function. $u$ is utility, $\mathbf{p}$ is the price vector.

26.22
$$x_i = \frac{m}{p_i}\left(\alpha_i + \sum_{j=1}^{n} \gamma_{ij} \ln p_j + \beta_i \ln\left(\frac{m}{\bar{p}}\right)\right),$$
where
$$\ln \bar{p} = \alpha_0 + \sum_{i=1}^{n} \alpha_i \ln p_i + \frac{1}{2} \sum_{i,j=1}^{n} \gamma_{ij} \ln p_i \ln p_j$$
with $\gamma_{ij} = \frac{1}{2}(\gamma_{ij}^* + \gamma_{ji}^*) = \gamma_{ji}$

Demand functions. Adding up: $\sum_{i=1}^{n} \alpha_i = 1$, $\sum_{i=1}^{n} \beta_i = 0$, $\sum_{i=1}^{n} \gamma_{ij} = 0$. Homogeneous provided $\sum_{i=1}^{n} \gamma_{ij} = 0$.

### Linear Expenditure System (LES)

26.23
$$u(\mathbf{x}) = \prod_{i=1}^{n} (x_i - \bar{x}_i)^{\beta_i}, \qquad \beta_i > 0$$

The LES utility function. If $\bar{x}_i = 0$ for all $i$, $u(\mathbf{x})$ is Cobb-Douglas.

26.24
$$x_i = \bar{x}_i + \frac{1}{p_i}\frac{\beta_i}{\beta}\left(m - \sum_{i=1}^{n} p_i \bar{x}_i\right)$$

Demand functions. $\beta = \sum_{i=1}^{n} \beta_i$.

26.25
$$v(\mathbf{p}, m) = \beta^{-\beta}\left(m - \sum_{i=1}^{n} p_i \bar{x}_i\right)^{\beta} \prod_{i=1}^{n} \left(\frac{\beta_i}{p_i}\right)^{\beta_i}$$

Indirect utility function.

26.26
$$e(\mathbf{p}, u) = \sum_{i=1}^{n} p_i \bar{x}_i + \frac{\beta u^{1/\beta}}{\left(\prod_{i=1}^{n} \left(\frac{\beta_i}{p_i}\right)^{\beta_i}\right)^{1/\beta}}$$

Expenditure function.

## Translog Indirect Utility function

26.27
$$\ln v = \alpha_0 + \sum_{i=1}^{n} \alpha_i \ln\left(\frac{p_i}{m}\right) + \frac{1}{2} \sum_{i,j=1}^{n} \beta_{ij}^* \ln\left(\frac{p_i}{m}\right) \ln\left(\frac{p_j}{m}\right)$$

The translog indirect utility function.

26.28
$$x_i = \frac{m}{p_i}\left(\frac{\alpha_i + \sum_{j=1}^{n} \beta_{ij} \ln(p_j/m)}{\sum_{i=1}^{n} \alpha_i + \sum_{i,j=1}^{n} \beta_{ij}^* \ln(p_i/m)}\right)$$

where $\beta_{ij} = \frac{1}{2}(\beta_{ij}^* + \beta_{ji}^*)$

Demand functions. Normalization:
$$\sum_{i=1}^{n} \alpha_i = -1.$$

## References

The basic reference is Varian (1989). For AIDS, see Deaton and Muellbauer (1980), for Translog, see Christensen, Jorgensen and Lau (1975).

## Chapter 27

# Topics from finance and growth theory

27.1 $S_t = S_{t-1} + rS_{t-1} = (1+r)S_{t-1}, \quad t = 1, 2, \ldots$

$S_t$ is the value of an asset in an account with interest rate $r$.

27.2 $S_t = S_0(1+r)^t$

The *compound amount* $S_t$ of a *principal* $S_0$ at the end of $t$ periods at the interest rate $r$ compounded at the end of each period.

27.3 Effective rate of interest $= (1 + \dfrac{r}{n})^n - 1$

The *effective rate of interest* (annual percentage rate) when interest is charged $n$ times a year at the rate of $r/n$ per period.

27.4 $A_t = \dfrac{R}{(1+r)^1} + \dfrac{R}{(1+r)^2} + \cdots + \dfrac{R}{(1+r)^t}$
$= R\dfrac{1-(1+r)^{-t}}{r}$

The *present value* $A_t$ of an *annuity* of $R$ per period for $t$ periods at the interest rate of $r$ per period.

27.5 $A = \dfrac{R}{(1+r)^1} + \dfrac{R}{(1+r)^2} + \cdots = \dfrac{R}{r}$

The *present value* $A$ of an *annuity* of $R$ per period for an infinite number of periods at the interest rate of $r$ per period.

27.6 $T = \dfrac{\ln\left(\dfrac{R}{R-rA}\right)}{\ln(1+r)}$

The number $T$ of periods required to pay off a loan of $A$ with periodic payment $R$ and interest rate $r$ per period.

27.7 $\quad S_t = (1+r)S_{t-1} + (y_t - x_t), \quad t = 1, 2, \ldots$

$S_t$ is the value of an asset in an account with deposits and withdrawals. $r$ is the interest rate, $y_t$ the deposits and $x_t$ the withdrawals in period $t$.

27.8 $\quad S_t = (1+r)^t S_0 + \sum_{k=1}^{t} (1+r)^{t-k}(y_k - x_k)$

The solution of equation (27.7)

27.9 $\quad (1+r)^{-t} S_t = S_0 + \sum_{k=1}^{t} (1+r)^{-k}(y_k - x_k)$

A reformulation of (27.8): The *present discounted value* (*PDV*) of the assets at time $t$ is equal to the initial assets $S_0$ plus the *PDV* of all future deposits $\sum_{k=1}^{t}(1+r)^{-k} y_k$, minus the *PDV* of all future withdrawals, $\sum_{k=1}^{t}(1+r)^{-k} x_k$.

27.10 $\quad S_t = (1+r_t)S_{t-1} + (y_t - x_t), \quad t = 1, 2, \ldots$

Generalization of (27.7) to the case with a variable interest rate, $r_t$.

27.11 $\quad D_k = \dfrac{1}{\prod_{s=1}^{k}(1+r_s)}$

The *discount factor* associated with (27.10).

27.12 $\quad R_k = \dfrac{D_k}{D_t} = \prod_{s=k+1}^{t}(1+r_s)$

The *interest factor* associated with (27.10).

27.13 $\quad S_t = R_0 S_0 + \sum_{k=1}^{t} R_k(y_k - x_k)$

The solution of (27.10). $R_k$ is defined in (27.12). (Generalizes (27.8).)

27.14 $\quad D_t S_t = S_0 + \sum_{k=1}^{t} D_k(y_k - x_k)$

Generalization of (27.9). The interpretation in (27.9) is still valid. Only the discount factor has changed.

27.15 $\quad a_0 + \dfrac{a_1}{1+r} + \dfrac{a_2}{(1+r)^2} + \cdots + \dfrac{a_n}{(1+r)^n} = 0$

$r$ is the *internal rate of return* of an investment project. Negative $a_i$ represents outlays, positive $a_i$ represents receipts.

| | | |
|---|---|---|
| 27.16 | If $a_0 < 0$ and $a_1, \ldots, a_n$ are all $\geq 0$, then (27.15) has a unique positive internal rate of return. | Consequence of Descartes' rule of signs (1.9). |
| 27.17 | $A_0 = a_0$, $A_1 = a_0 + a_1$, $A_2 = a_0 + a_1 + a_2$, $\ldots$, $A_n = a_0 + a_1 + \cdots + a_n$ | The *accumulated cash flow* associated with (27.15). |
| 27.18 | If $A_n \neq 0$, and the sequence $A_0, A_1, \ldots, A_n$ changes sign only once, then (27.15) has a unique internal rate of return. | Norstrøm's rule. |
| 27.19 | • $X(t) = F(K(t), L(t))$<br>• $\dot{K}(t) = sX(t)$<br>• $L(t) = L_0 e^{\lambda t}$ | A growth model. $X(t)$ is national income, $K(t)$ is capital, and $L(t)$ is the labor force at time $t$. $F$ is a production function. $s$ (the rate of savings), $\lambda$, and $L_0$ are positive constants. |
| 27.20 | If $F$ is homogeneous of degree 1, $k(t) = K(t)/L(t)$ is capital per worker, and $f(k) = F(k, 1)$, then (27.19) reduces to<br>$\dot{k} = sf(k) - \lambda k$, $k(0)$ is given | Solow's growth model. |
| 27.21 | If $\lambda/s < f'(0) < \infty$, $f'(k) \to 0$ as $k \to \infty$, and $f''(k) \leq 0$ for all $k \geq 0$, then the equation in (27.20) has a unique solution on $[0, \infty)$. The solution $k^*$, defined by<br>$sf(k^*) = \lambda k^*$<br>is a stable equilibrium state. | The existence and uniqueness of a solution on $[0, \infty)$ follows from (10.42). |
| 27.22 | $\max \int_0^T U(f(K(t)) - \dot{K}(t)) e^{-rt} \, dt$<br>$K(0) = K_0$, $K(T) \geq K_1$ | A standard problem in growth theory. $U$ is a utility function, $K(t)$ is the capital stock at time $t$, $f(K)$ is the production function, $r$ is the discount factor, and $T$ is the planning horizon. (The Ramsey model.) |
| 27.23 | $\ddot{K} - f'(K)\dot{K} + \dfrac{U'(C)}{U''(C)}(r - f'(K)) = 0$ | The Euler equation for problem (27.22). |

27.24 $\dfrac{\dot{C}}{C} = \dfrac{r - f'(K)}{\check{w}}$, where

$\check{w} = El_C U'(C) = CU''(C)/U'(C)$

Necessary condition for the solution of (27.22).

## References

For compound interest formulas see Goldberg (1961). For (27.18) see Norstrøm (1972). For growth theory, see Burmeister and Dobell (1970) or Wan (1971).

# Chapter 28

# Risk and risk aversion theory

28.1 $\quad R_A = -\dfrac{u''(y)}{u'(y)}\quad$ *Absolute risk aversion.* $u(y)$: utility function, $y$: income.

28.2 $\quad R_R = yR_A = -\dfrac{yu''(y)}{u'(y)}\quad$ *Relative risk aversion.*

28.3 $\quad \begin{aligned} E\left(u(y+z+\pi)\right) &= E\left(u(y)\right) \\ \pi &\approx -\dfrac{u''(y)}{u'(y)}\dfrac{\sigma^2}{2} = R_A\dfrac{\sigma^2}{2} \end{aligned}\quad$ *Arrow-Pratt risk premium.* $z$: mean zero risky prospect. $\pi$: risk premium. $\sigma^2$: variance of $z$. $E(\ )$ is expectation.

28.4 $\quad\begin{aligned} u(y) &= y^\alpha,\ \alpha\in(0,1),\ R_A = (1-\alpha)y^{-1} \\ u(y) &= \ln y,\ R_A = y^{-1} \\ u(y) &= 1 - e^{-\lambda y},\ \lambda > 0,\ R_A = \lambda \\ u(y) &= y - by^2,\ b>0,\ R_A = \dfrac{2b}{1-2by} \end{aligned}\quad$ Risk measures for four utility functions.

28.5 If $F$ and $G$ are cumulative distribution functions (CDF) of random incomes,

$F$ first degree stochastically dominates $G$
$\iff G(X) \geq F(X)$ for all $X \in I$

*First degree stochastic dominance.* $I$ is a closed interval $[X_1, X_2]$. For $X \leq X_1$, $F(X) = G(X) = 0$ and for $X \geq X_2$, $F(X) = G(X) = 1$.

28.6 $\quad F\ \text{FSD}\ G \iff \begin{cases} E_F\left(u(X)\right) \geq E_G\left(u(X)\right) \\ \text{for all increasing } u(X) \end{cases}\quad$ An important result. FSD means "first degree stochastically dominates". $E_F$ is expectation with CDF $F$. $u(X)$ is utility.

28.7 $T(X) = \int_{X_1}^{X} (G(x) - F(x))\, dx$ — A definition.

28.8 $F$ second degree stochastically dominates $G$ $\iff T(X) \geq 0$ for all $X \in I$

*Second degree stochastic dominance* (SSD), $I = [X_1, X_2]$. Note that FSD $\Rightarrow$ SSD.

28.9 $F$ SSD $G \iff \begin{cases} E_F(u(X)) \geq E_G(u(X)) \\ \text{for all increasing and} \\ \text{concave } u(X) \end{cases}$

Hadar-Russell's theorem. Every risk averter prefers $F$ to $G$ if and only if $F$ SSD $G$

28.10 $Y$ is $X$ + noise $\iff Y =_d X + Z$, where $E(Z|X) = 0$ for all $X$ and some $Z$.

$=_d$ means: "has same distribution"

28.11 The following statements are equivalent:
- $T(X_2) = 0$ and $T(X) \geq 0$ for all $X \in I$
- $G$ is distributed as $F$ plus noise.
- $F$ and $G$ have the same mean and every risk averter prefers $F$ to $G$.

Rothschild-Stiglitz's theorem.

## References

See Huang and Litzenberger (1988), Hadar and Russel (1969), and Rothschild and Stiglitz (1970).

# Chapter 29

# Finance and stochastic calculus

29.1
$$E[R_i] - R = \beta_i(E[R_m] - R), \text{ where}$$
$$\beta_i = \frac{\text{corr}(R_i, R_m)\sigma_i}{\sigma_m} = \frac{\text{cov}(R_i, R_m)}{\sigma_m^2}$$

*Capital asset pricing model.* $R_i$: rate of return on asset i. $E[R_k]$ is the expected value of $R_k$. $R$: rate of return on safe asset. $R_m$: market rate of return. $\sigma_i$: standard deviation of $R_i$.

29.2
$$E(R_i) - R = \frac{\beta_{ic}}{\beta_{mc}}(E(R_m) - R), \text{ where}$$
$$\beta_{jc} = \frac{\text{cov}(R_j, d\ln C)}{\text{var}(d\ln C)}, \quad j = i \text{ or } m$$

*Single consumption $\beta$ asset pricing equation.* $C$: consumption. $R_m$: return on any portfolio. $d\ln C$ is the stochastic logarithmic differential. (See (29.8))

29.3
$$c(S, X, R, v, t^* - t) =$$
$$SN(d_1) - XN(d_2)e^{-R(t^*-t)}, \text{ where}$$
$N(x)$ is the standard normal $CDF$ at $x$,
$$d_1 = \frac{\ln(S/X) + (R + \tfrac{1}{2}v^2)(t^* - t)}{v(t^* - t)^{1/2}}$$
$$d_2 = d_1 - v(t^* - t)^{1/2}$$

*Black and Schole's option pricing model.* (European or American call option on non-dividend paying stock.) $S$: underlying stock price. $c(S,t)$ = the value of the option on $S$ at time $t$. $dS/S = \alpha dt + vdZ$, where $Z$ is a (standard) Brownian motion. $t^*$: expiration date. $R$: interest rate. $X$: exercise price.

| | | |
|---|---|---|
| 29.4 | $$c(S, X, R, v, t^* - t, \delta)$$ $$= SN(d_1)e^{-\delta(t^*-t)} - XN(d_2)e^{-R(t^*-t)}$$ where $N(x)$ is the standard normal $CDF$ at $x$, $$d_1 = \frac{\ln(S/X) + (R - \delta + \tfrac{1}{2}v^2)(t^* - t)}{v(t^* - t)^{1/2}}$$ $$d_2 = d_1 - v(t^* - t)^{1/2}$$ | European call option on an asset which earns a below-equilibrium rate of return (or on a stock which pays a continuous proportional dividend.) $\delta$ = rate of return shortfall = marginal net relative convenience yield = continuous dividend yield. If $\delta = 0$, the model reduces to the Black and Schole's model. |
| 29.5 | $X_t = X_0 + \int_0^t u(s,\omega)\,ds + \int_0^t v(s,\omega)\,dB_s$ where $P[\int_0^t v(s,\omega)^2\,ds < \infty \text{ for all } t \geq 0] = 1$ and $P[\int_0^t |u(s,\omega)|\,ds < \infty \text{ for all } t \geq 0] = 1$, Both $u$ and $v$ are adapted to the filtration $\{\mathcal{F}_t\}$, where $B_t$ is an $\mathcal{F}_t$-Brownian motion. | $X_t$ is by definition a one dimensional *stochastic integral*. |
| 29.6 | $dX_t = u\,dt + v\,dB_t$ | A differential form of (29.5). |
| 29.7 | If $dX_t = u\,dt + v\,dB_t$ and $Y_t = g(X_t)$, where $g \in C^2$, then $$dY_t = \left(g'(X_t) + \tfrac{1}{2}g''(X_t)\right)dt + g'(X_t)v\,dB_t$$ | *Itô's formula* (one-dimensional). |
| 29.8 | $dt \cdot dt = dt \cdot dB_t = dB_t \cdot dt = 0,\ dB_t \cdot dB_t = dt$ | Useful relations. |
| 29.9 | $$d\ln X_t = \left(\frac{u}{X_t} - \frac{v^2}{2X_t^2}\right)dt + \frac{v}{X_t}dB_t$$ $$de^{X_t} = \left(e^{X_t}u + \frac{1}{2}e^{X_t}v^2\right)dt + e^{X_t}v\,dB_t$$ | Two special cases of (29.7). |
| 29.10 | $$\begin{pmatrix} dX_1 \\ \vdots \\ dX_n \end{pmatrix} = \begin{pmatrix} u_1 \\ \vdots \\ u_n \end{pmatrix} dt + \begin{pmatrix} v_{11} & \cdots & v_{1m} \\ \vdots & & \vdots \\ v_{n1} & \cdots & v_{nm} \end{pmatrix} \begin{pmatrix} dB_1 \\ \vdots \\ dB_m \end{pmatrix}$$ | Vector version of (29.6), where $B_1, \ldots, B_m$ are $m$ independent one-dimensional Brownian motions. |

29.11 If $\mathbf{Y} = (Y_1, \ldots, Y_k) = \mathbf{g}(t, \mathbf{X})$, where $\mathbf{g} = (g_1, \ldots, g_k)$ is $C^2$, then for $r = 1, \ldots, k$

$$dY_r = \frac{\partial g_r(t, \mathbf{X})}{\partial t} dt + \sum_{i=1}^{n} \frac{\partial g_r(t, \mathbf{X})}{\partial x_i} dX_i$$

$$+ \frac{1}{2} \sum_{i,j=1}^{n} \frac{\partial^2 g_r(t, \mathbf{X})}{\partial x_i \partial x_j} dX_i dX_j$$

where $dt \cdot dt = dt \cdot dB_i = 0$, $dB_i \cdot dB_j = dt$ if $i \neq j$, 0 if $i = j$.

An $n$-dimensional *Itô formula*.

29.12 $J(t, x) = \max_u E^{t,x} \int_t^T e^{-rs} W(s, X_s, u_s) ds$,

where $T$ is fixed, $u_s \in U$, $U$ a fixed interval, and

$$dX_t = b(t, X_t, u_t) dt + \sigma(t, X_t, u_t) dB_t$$

A *stochastic control problem*. $J$ is the value function, $u_t$ is the control. $E^{t,x}$ is expectation when the initial condition is $X_t = x$.

29.13 $-J'_t(t, x) = \max_{u \in U} [W(t, x, u)$
$+ J'_x(t, x) b(t, x, u) + \frac{1}{2} J''_{xx}(t, x) (\sigma(t, x, u))^2]$

The *Hamilton-Jacobi-Bellman equation*. A necessary condition for optimality in (29.12).

## References

For (29.1) and (29.2) see Sharpe (1964). For (29.3) see Black and Scholes (1973). For (29.4) see McDonald and Siegel (1984). For stochastic calculus and stochastic control theory, see Øksendal (1989), Fleming and Rishel (1975), or Karatzas and Shreve (1988). For Itô formulas for general semimartingales (describing processes with jumps), see Protter (1990).

# Chapter 30

# Non-cooperative game theory

30.1 In an $n$-person game we assign to each player $i$ a *strategy set* $S_i$ and a a real valued pure strategy *payoff function* $u_i(s_1,\ldots,s_n)$, where $s_i \in S_i$ for $i = 1,\ldots,n$.

An $n$-person game.

30.2 A *strategy profile* $(s_1^*,\ldots,s_n^*)$ for an $n$-person game is a *pure strategy Nash equilibrium* if for all $s_i \in S_i$ and all $i = 1,\ldots,n$,
$$u_i(s_1^*,\ldots,s_n^*) \geq u_i(s_1^*,\ldots,s_{i-1}^*, s_i, s_{i+1}^*,\ldots s_n^*).$$

Definition of a pure strategy Nash equilibrium for an $n$-person game.

30.3 If for all $i = 1,\ldots,n$, $S_i$ is non-empty, compact, and convex, and $u_i(s_1,\ldots,s_n)$ is continuous in $S_1 \times \cdots \times S_n$ and quasi-concave in its $i$th variable, then the game has a pure strategy Nash equilibrium.

Sufficient conditions for the existence of a pure strategy Nash equilibrium. (Typically there will be several Nash equilibria.)

30.4 Consider a *finite* game in which player $i$'s strategy set, $S_i$, has $k_i$ elements, $s_i^{j_i}$, $j_i = 1,\ldots,k_i$. A *mixed strategy* for player $i$ is a vector of probability weights, $\sigma_i = (\sigma_{i1},\ldots,\sigma_{ik},\ldots,\sigma_{ik_i})$, i.e. for all $k$, $\sigma_{ik} \geq 0$ and $\sum_{k=1}^{k_i} \sigma_{ik} = 1$. ($\sigma_{ik}$ is the probability that $i$ choooses the $k$'th strategy in $S_i$.) A mixed strategy profile $\bar{\sigma} = (\bar{\sigma}_1,\ldots,\bar{\sigma}_n)$ is a *Nash equilibrium* if for all $i$ and each $\sigma_i$,
$$u^i(\bar{\sigma}) \geq u^i(\bar{\sigma}_1,\ldots,\bar{\sigma}_{i-1},\sigma_i,\bar{\sigma}_{i+1},\ldots,\bar{\sigma}_n),$$
where
$$u^i(\sigma_1,\ldots,\sigma_n) = \sum_{j_1}\cdots\sum_{j_n} \sigma_{1j_1}\cdots\sigma_{1j_n} u_i(s_1^{j_1},\ldots,s_n^{j_n})$$
Here $j_i$ runs through $\{1,\ldots,k_i\}$.
Every finite game has a Nash equilibrium.

Definition and existence of a (mixed strategy) Nash equilibrium for a finite game. $u^i(\sigma_1,\ldots,\sigma_n)$ is the expected utility of player $i$ when all players play mixed strategies $(\sigma_1,\ldots,\sigma_n)$.

30.5 Let $\sigma_i^k$ denote the mixed strategy for $i$ that assigns probability one to $i$'s $k$th strategy. A mixed strategy profile is a Nash equilibrium if and only if for all $i$ and for all $1 \leq k \leq k_i$,
$$u^i(\bar{\sigma}) \geq u^i(\bar{\sigma}_1, \ldots, \bar{\sigma}_{i-1}, \sigma_i^k, \bar{\sigma}_{i+1} \bar{\sigma}_n)$$

An equivalent definition of a (mixed strategy) Nash equilibrium.

30.6 A two-person game where the players I and II have $m$ and $n$ (pure) strategies respectively, can be represented by the two payoff-matrices
$$\begin{pmatrix} a_{11} & \cdots & a_{1n} \\ a_{21} & \cdots & a_{2n} \\ \vdots & & \vdots \\ a_{m1} & \cdots & a_{mn} \end{pmatrix}, \begin{pmatrix} b_{11} & \cdots & b_{1n} \\ b_{21} & \cdots & b_{2n} \\ \vdots & & \vdots \\ b_{m1} & \cdots & b_{mn} \end{pmatrix}$$

$a_{ij}$ ($b_{ij}$) is the payoff to player I (II) in von Neumann-Morgenstern units when the players play their pure strategies $i$ and $j$, respectively.

30.7 Let $(a_{ij})$ and $(b_{ij})$ be the matrices in (30.6). Then there exist mixed strategies $\mathbf{p}^* = (p_1^*, \ldots, p_m^*) \in \mathcal{P}_I$ and $\mathbf{q}^* = (q_1^*, \ldots, q_n^*) \in \mathcal{P}_{II}$ such that
- for all $\mathbf{p} \in \mathcal{P}_I$,
$$\sum_{j=1}^n \sum_{i=1}^m a_{ij} p_i q_j^* \leq \sum_{j=1}^n \sum_{i=1}^m a_{ij} p_i^* q_j^*$$
- for all $\mathbf{q} \in \mathcal{P}_{II}$,
$$\sum_{j=1}^n \sum_{i=1}^m b_{ij} p_i^* q_j \leq \sum_{j=1}^n \sum_{i=1}^m b_{ij} p_i^* q_j^*$$

(30.4) specialized to a two-person game. (If player II chooses $\mathbf{q}^*$, the best I can do is to choose $\mathbf{p}^*$. If player I chooses $\mathbf{p}^*$, the best II can do is to choose $\mathbf{q}^*$).

30.8 A two-person *zero sum* game where the players have $n$ and $m$ (pure) strategies respectively, can be represented by the payoff matrix
$$\mathbf{A} = \begin{pmatrix} a_{11} & a_{12} & \cdots & a_{1n} \\ a_{21} & a_{22} & \cdots & a_{2n} \\ \vdots & \vdots & & \vdots \\ a_{m1} & a_{m2} & \cdots & a_{mn} \end{pmatrix}$$

Since the game is zero sum, $a_{ij}$ has the interpretation of *money* received by player I from player II if they play the (pure) strategies $i$ and $j$, respectively.

30.9 Let $\mathbf{A}$ be the matrix in (30.8). There exist a unique number $v$ and mixed strategies $\mathbf{p}^* = (p_1^*, \ldots, p_m^*) \in \mathcal{P}_I$ and $\mathbf{q}^* = (q_1^*, \ldots, q_n^*) \in \mathcal{P}_{II}$ such that
- $\sum_{i=1}^m a_{ij} p_i^* \geq v$ for all $j = 1, \ldots, n$
- $\sum_{j=1}^n a_{ij} q_j^* \leq v$ for all $i = 1, \ldots, m$

The classical *minimax theorem* for two-person zero sum games.

| | | |
|---|---|---|
| 30.10 | The number $v$ (30.9) is called the *value* of the game, and $$v = \sum_{j=1}^{n}\sum_{i=1}^{m} a_{ij} p_i^* q_j^*$$ | $v$ is the expected profit for player I when he chooses strategy $\mathbf{p}^*$ and II chooses $\mathbf{q}^*$. |
| 30.11 | Assume that $(\mathbf{p}^*, \mathbf{q}^*)$ and $(\mathbf{p}^{**}, \mathbf{q}^{**})$ are Nash equilibria in the game (30.8). Then $(\mathbf{p}^*, \mathbf{q}^{**})$ and $(\mathbf{p}^{**}, \mathbf{q}^*)$ are also equilibrium strategy profiles. | The *rectangular* or *exchangeability* property. |

## References

A standard reference is Friedman (1986). See also Kreps (1990).

# Chapter 31

# Statistical concepts

31.1    $P(A) \in [0,1]$ is the probability that a random event $A$ occurs.   |   A definition.

31.2    $P(A \mid B) = \dfrac{P(A \cap B)}{P(B)}$ is the *conditional probability* that event $A$ will occur given that $B$ has occurred.   |   Definition of *conditional probability*.

31.3    $A$ and $B$ are *stochastically independent* if $P(A \mid B) = P(A)$.   |   Definition of *stochastic independence*.

31.4    If $A$ and $B$ are *stochastically independent*, then $P(A \cap B) = P(A)P(B)$, and $P(B \mid A) = P(B)$.   |   Useful facts.

31.5    $P(A \mid B) = \dfrac{P(B \mid A) \cdot P(A)}{P(B)}$   |   Bayes' rule.

31.6    $P(A_i \mid B) = \dfrac{P(B \mid A_i) \cdot P(A_i)}{\sum\limits_{j=1}^{n} P(B \mid A_j) \cdot P(A_j)}$   |   Bayes' rule. $A_1, \ldots, A_n$ are disjoint, $\sum_{i=1}^{n} P(A_i) = P(A) = 1$, where $A = \bigcup\limits_{i=1}^{n} A_i$ is the sample space.

31.7    If $F(x) = P(X \leq x)$ is a cumulative distribution function (CDF) for the random variable $X$, then $f(x) = dF/dx$ (if it exists) is the *probability density function* of $X$.   |   Definition of the *probability density function*.

31.8    $E(X) = \sum\limits_{i=1}^{n} \pi_i x_i$,   $\sum\limits_{i=1}^{n} \pi_i = 1$   |   Definition of *expectation* for a discrete random variable $X$, which takes the value $x_i$ with probability $\pi_i$.

31.9 $\quad E(X) = \int\limits_{-\infty}^{\infty} x f(x)\, dx$ 
           Definition of *expectation* for a continuous distribution.

31.10 $\quad E(X) = \int\limits_{-\infty}^{\infty} x\, dF$ 
           General definition of expectation if the distribution function $F$ is not differentiable, using a Lebesgue-Stieltjes integral.

31.11 $\quad \begin{aligned} &E(a_1 X_1 + \cdots + a_n X_n + b) = \\ &a_1 E X_1 + \cdots + a_n E X_n + b \end{aligned}$ 
           $X_1, \ldots, X_n$ are random variables and $a_1, \ldots, a_n$, $b$ are real numbers.

31.12 $\quad \mathrm{var}(X) = E((X - EX)(X - EX))$ 
           Definition of *variance*.

31.13 $\quad \mathrm{var}(X) = E(X^2) - (EX)^2$ 
           Another expression for the variance.

31.14 $\quad \sigma = \sqrt{\mathrm{var}(X)}$ 
           The *standard deviation* of $X$.

31.15 $\quad \mathrm{var}(aX + b) = a^2 \mathrm{var}(X)$ 
           $a$ and $b$ are real numbers.

31.16 $\quad \mathrm{cov}(X, Y) = E((X - EX)(Y - EY))$ 
           Definition of *covariance*.

31.17 $\quad \mathrm{cov}(X, Y) = E(XY) - E(X)E(Y)$ 
           A useful fact.

31.18 $\quad$ If $\mathrm{cov}(X, Y) = 0$, $X$ and $Y$ are called *uncorrelated*. 
           A definition.

31.19 $\quad E(XY) = E(X)E(Y)$ if $X$ and $Y$ are uncorrelated. 
           Follows from (31.17) and (31.18).

31.20 $\quad$ If $X$ and $Y$ are stochastically independent, then $\mathrm{cov}(X, Y) = 0$. 
           The converse is not true.

31.21 $\quad \mathrm{var}(X \pm Y) = \mathrm{var}(X) + \mathrm{var}(Y) \pm 2\,\mathrm{cov}(X, Y)$ 
           The variance of a sum/difference of two stochastic variables.

31.22 $\quad \mathrm{var}\left(\sum\limits_{i=1}^{n} a_i X_i\right) = \sum\limits_{i=1}^{n} a_i^2 \mathrm{var}(X_i) + 2 \sum\limits_{i>j}^{n} \sum\limits_{j=1}^{n-1} \mathrm{cov}(X_i, X_j)$ 
           The variance of a linear combination of stochastic variables.

| | | |
|---|---|---|
| 31.23 | $\text{var}(\sum_{i=1}^{n} a_i X_i) = \sum_{i=1}^{n} a_i^2 \text{var}(X_i)$ | Formula (31.22) when $X_1, \ldots, X_n$ are uncorrelated. |
| 31.24 | $\text{corr}(X, Y) = \dfrac{\text{cov}(X, Y)}{\sqrt{\text{var}(X)\text{var}(Y)}} \in [-1, 1]$ | Definition of the *correlation coefficient* as a normalized covariance. |
| 31.25 | $P(|X - E(X)| \geq \lambda) \leq \text{var}(X)/\lambda^2$ for all $\lambda > 0$ | *Chebychev's inequality.* |
| 31.26 | If $E(\hat{\theta}) = \theta$, then $\hat{\theta}$ is called an *unbiased estimator* of $\theta$. | Definition of an unbiased estimator. |
| 31.27 | When $\hat{\theta}$ is not unbiased, then $b = E(\hat{\theta}) - \theta$ is the bias of $\hat{\theta}$. | Definition of *bias*. |
| 31.28 | $\text{MSE}(\hat{\theta}) = E(\hat{\theta} - \theta)^2 = \text{Var}(\hat{\theta}) + b^2$ | Definition of *mean square error*. |
| 31.29 | $\text{plim}\,\hat{\theta}_T = \theta$ means that for every $\varepsilon > 0$ $$\lim_{T \to \infty} P(|\hat{\theta}_T - \theta| < \varepsilon) = 1$$ | Definition of a *probability limit*. The estimator $\hat{\theta}_T$ is a function of $T$ observations. |
| 31.30 | $\hat{\theta}$ is a *consistent* estimator of $\theta$ if $p\lim \hat{\theta}_T = \theta$ | Definition of *consistency*. |
| 31.31 | $\hat{\theta}$ is *asymptotically unbiased* if $$\lim_{T \to \infty} E(\hat{\theta}_T) = \theta.$$ | Definition of *asymptotically unbiased* estimator. |
| 31.32 | If $f(x, y)$ is the joint density of $X$ and $Y$, then $$g(x) = \int_{-\infty}^{\infty} f(x, y)\, dy$$ is the *marginal density of $X$*; $$h(y) = \int_{-\infty}^{\infty} f(x, y)\, dx$$ is the *marginal density of $Y$*. | Definition of *marginal densities*. |
| 31.33 | $f(x \mid y) = \dfrac{f(x, y)}{h(y)}, \quad h(y) \neq 0$ | Definition of *conditional density*. |
| 31.34 | The random variables $X$ and $Y$ are *stochastically independent* if $f(x \mid y) = g(x)$. Then $f(x, y) = g(x) h(y)$. | A definition and a result associated with (31.31) and (31.32). |

| | | | |
|---|---|---|---|
| 31.35 | $H_0$ | Null hypothesis (e.g. $\theta \leq 0$) | Definitions for *statistical testing*. |
| | $H_1$ | Alternate hypothesis (e.g. $\theta > 0$) | |
| | $T$ | Test statistics | |
| | $C$ | Critical region | |
| | $\theta$ | an unknown parameter | |

31.36    A test: Reject $H_0$ in favour of $H_1$ if $T \subset C$.   |   A test.

31.37    The power function of a test is $P(\text{reject } H_0|\theta)$, $\theta \in \{\theta : H_0 \text{ false}\}$   |   Definition of the *power* of a test.

31.38    To reject $H_0$ when $H_0$ is true is called a type I error.
To accept $H_0$ when $H_1$ is true is called a type II error.   |   Type I and II errors.

31.39    $\alpha$-level of significance: The least $\alpha$ such that $P(\text{type I error}) \leq \alpha$ for all $\theta$ satisfying $H_0$.   |   The *$\alpha$-level of significance* of a test.

## References

See e.g. Judge et al (1988), or Hogg and Craig (1978).

# Chapter 32

## Statistical distributions. Least squares

32.1 $$f(x;n,p) = \binom{n}{x} p^x (1-p)^{n-x}$$
$x = 0, 1, \ldots, \quad n = 1, 2, \ldots, \quad p \in (0, 1)$

Mean: $E(X) = np$.
Variance: $\operatorname{var}(X) = np(1-p)$.

The *binomial distribution*. $f(x; n, p)$ is the probability for an event to occur exactly $x$ times in $n$ independent observations, when the probability of the event is $p$ at each observation.

32.2 $$f(x_1, \ldots, x_{k-1}; n, p_1, \ldots, p_{k-1}) = \frac{n!}{x_1! \cdots x_k!} p_1^{x_1} \cdots p_k^{x_k},$$
$x_1 + \cdots + x_k = n, \quad p_1 + \cdots + p_k = 1,$
$x_j \in \{0, 1, \ldots, n\}, \quad p_j \in (0, 1), \quad j = 1, \ldots, k$

Mean of $X_j$: $E(X_j) = np_j$.
Variance of $X_j$: $\operatorname{var}(X_j) = np_j(1-p_j)$.
The covariance of $X_j X_r$:
$\operatorname{cov}(X_j, X_r) = -np_j p_r \quad r, j = 1, \ldots, n, \ r \neq j$

The *multinomial distribution*. $f(\ )$ is the probability for $k$ events $A_1, \ldots, A_k$ to occur exactly $x_1, \ldots, x_k$ times in $n$ independent observations, when the probabilities of the events are $p_1, \ldots, p_k$.

32.3 $$f(x; N, M, n) = \frac{\binom{M}{x} \binom{N-M}{n-x}}{\binom{N}{n}}$$
$x = 0, 1, \ldots, \quad n = 1, 2, \ldots, \quad n \leq N$
$0 \leq x \leq M, \ 0 \leq n - x \leq N - M$

Mean: $E(X) = nM/N$.
Variance: $\operatorname{var}(X) = np(1-p)(N-n)/(N-1)$,
where $p = M/N$.

The *hypergeometric distribution*. Given a collection of $N$ objects, where $M$ objects have a certain characteristic and $N - M$ do not have it. Suppose we pick at random $n$ objects from the collection. $f(x; N, M, n)$ is then the probability that $x$ objects have the characteristic and $n - x$ do not have it.

32.4 $f(x;\lambda) = e^{-\lambda}\dfrac{\lambda^x}{x!}, \quad \lambda > 0, \quad x = 0, 1, 2, \ldots$

Mean: $E(X) = \lambda$.
Variance: $\mathrm{var}(X) = \lambda$.

*Poisson distribution.*

32.5 $f(x;\mu,\sigma) = \dfrac{1}{\sigma\sqrt{2\pi}}\, e^{-(x-\mu)^2/2\sigma^2}, \quad \sigma > 0$

Mean: $E(X) = \mu$.
Variance: $\mathrm{var}(X) = \sigma^2$.

*Normal distribution.*

32.6 $\phi(x) = \dfrac{1}{\sqrt{2\pi}}\, e^{-x^2/2}$

Mean: $E(X) = 0$.
Variance: $\mathrm{var}(X) = 1$.

*The standardized normal distribution.*

32.7 $f(x;\mu,\sigma) = \dfrac{1}{x\sqrt{2\pi\sigma^2}} e^{-(\log x - \mu)^2/2\sigma^2}$

$\sigma > 0, \; y > 0$

Mean: $E(X) = e^{\mu + (\sigma^2/2)}$.
Variance: $\mathrm{var}(X) = e^{2\mu}(e^{2\sigma^2} - e^{\sigma^2})$.

*Lognormal distribution.*

32.8 $f(x,y;\mu,\eta,\sigma,\tau,\rho) = \dfrac{e^{-Q}}{\sigma\tau 2\pi\sqrt{1-\rho^2}}, \quad$ where

$Q = \dfrac{-\left(\left(\frac{x-\mu}{\sigma}\right)^2 - 2\rho\frac{(x-\mu)(y-\eta)}{\sigma\tau} + \left(\frac{y-\eta}{\tau}\right)^2\right)}{2(1-\rho^2)}$

$x, y, \mu, \eta \in (-\infty, \infty), \; \sigma > 0, \; \tau > 0, \; |\rho| < 1$

Mean: $E(X) = \mu \quad E(Y) = \eta$,
Variance: $\mathrm{var}(X) = \sigma^2 \quad \mathrm{var}(Y) = \tau^2$.
Covariance of $XY$: $\mathrm{cov}(X,Y) = \rho\sigma\tau$.

*The binormal distribution.*

32.9 $f(\mathbf{x};\mu,\boldsymbol{\Sigma}) = \dfrac{1}{(2\pi)^{k/2}\sqrt{|\boldsymbol{\Sigma}|}} e^{-\frac{1}{2}(\mathbf{x}-\mu)\boldsymbol{\Sigma}^{-1}(x-\mu)'}$

$\boldsymbol{\Sigma} = (\sigma_{ij})$ is a symmetric, pos. definite matrix,
$\mathbf{x} = (x_1, \ldots, x_k), \; \mu = (\mu_1, \ldots, \mu_k)$.

Mean: $E(x_i) = \mu_i$.
Variance: $\mathrm{var}(x_i) = \sigma_{ii}$.
Covariance: $\mathrm{cov}(x_i, x_j) = \sigma_{ij}$.

*Multidimensional normal distribution.*

32.10 $f(x;\alpha) = \begin{cases} \alpha e^{-\alpha x}, & x > 0 \\ 0, & x \le 0 \end{cases} \quad (\alpha > 0)$

Mean: $E(X) = 1/\lambda$.
Variance: $\mathrm{var}(X) = 1/\lambda^2$.

*Exponential distribution.*

32.11  $f(x;\alpha,\beta) = \dfrac{1}{\beta-\alpha}, \quad \alpha < \beta,\ \alpha < x < \beta$

Mean: $E(X) = (\alpha+\beta)/2$.
Variance: $\text{var}(X) = (\beta-\alpha)^2/12$.

*Uniform rectangular distribution.*

32.12  $f(x;p) = p(1-p)^x, \quad p > 0,\ x = 0,1,2,\ldots$

Mean: $E(X) = (1-p)/p$.
Variance: $\text{var}(X) = (1-p)/p^2$.

*Geometric distribution.*

32.13  $f(x;p) = \dfrac{x^{p-1} e^{-x/\beta}}{\beta^p \Gamma(p)}, \quad p,\beta > 0$

Mean: $E(X) = \beta p$.
Variance: $\text{var}(X) = \beta^2 p$.

*Gamma distribution.*

32.14  $f(x;p,q) = \dfrac{x^{p-1}(1-x)^{q-1}}{B(p,q)}, \quad p,q > 0,\ x > 0$

Mean: $E(X) = \dfrac{p}{p+q}$.
Variance: $\text{var}(X) = \dfrac{pq}{(p+q)^2(p+q+1)}$.

*Beta distribution. B is the Beta function defined in (8.56)*

32.15  $f(z;\nu) = \dfrac{z^{\nu/2-1} e^{-z/2}}{2^{\nu/2}\Gamma(\nu/2)}, \quad z > 0,\ \nu = 1,2,\ldots$

Mean: $E(X) = \dfrac{p}{p+q}$.
Variance: $\text{var}(X) = \dfrac{pq}{(p+q)^2(p+q+1)}$.

*Chi-square distribution with $\nu$ degrees of freedom. $\Gamma$ is the gamma function defined in (8.51).*

32.16  $f(t;\nu) = \dfrac{\Gamma((\nu+1)/2)}{\sqrt{\nu\pi}\,\Gamma(\nu/2)}\left(1+\dfrac{t^2}{\nu}\right)^{-\frac{\nu+1}{2}}, \quad \nu = 1,2,\ldots$

Mean: $E(X) = 0,\ \nu > 1$
(does not exist for $\nu = 1$).
Variance: $\text{var}(X) = \nu/(\nu-2),\ \nu \geq 3$
(does not exist for $\nu = 1, 2$).

*Student t-distribution with $\nu$ degrees of freedom.*

32.17  $\dot{g}(x;\nu_1,\nu_2) = \dfrac{\nu_1^{(\nu_1/2)} \nu_2^{(\nu_2/2)} x^{\nu_1/2-1}}{B(\nu_1/2,\nu_2/2)(\nu_2+\nu_1 x)^{(\nu_1+\nu_2)/2}}$

$x > 0,\ \nu_1,\nu_2 = 1,2,\ldots$

Mean: $E(X) = \nu_2/(\nu_2-2),\ \nu_2 \geq 3$
(does not exist for $\nu_2 = 1, 2$).
Variance: $\text{var}(X) = \dfrac{2\nu_2^2(\nu_1+\nu_2-2)}{\nu_1(\nu_2-2)^2(\nu_2-4)}$,
$\nu_2 > 4$, (does not exist for $\nu_2 \leq 4$).

*F-distribution. B is the Beta function defined in (8.56). $\nu_1, \nu_2$ are the degrees of freedom for the numerator and denominator resp.*

32.18 The straight line which best fits $n$ data points $(x_1, y_1), (x_2, y_2), \ldots, (x_n, y_n)$, in the sense that the sum of the square deviations,

$$\sum_{i=1}^{n} \left(y_i - (ax_i + b)\right)^2,$$

is minimal, is given by $y - \bar{y} = \hat{a}(x - \bar{x})$, where $\bar{x} = \frac{1}{n}\sum_{i=1}^{n} x_i$, $\bar{y} = \frac{1}{n}\sum_{i=1}^{n} y_i$, and

$$\hat{a} = \frac{\sum_{i=1}^{n}(x_i - \bar{x})(y_i - \bar{y})}{\sum_{i=1}^{n}(x_i - \bar{x})^2}$$

The *method of least squares.*

Given $n$ observations $(x_{i1}, \ldots, x_{ik})$, $i = 1, \ldots, n$, of $k$ quantities $x_1, \ldots, x_k$, and $n$ observations $y_1, \ldots, y_n$ of a corresponding quantity $y$. Define

$$\mathbf{X} = \begin{pmatrix} 1 & x_{11} & x_{12} & \cdots & x_{1k} \\ 1 & x_{21} & x_{22} & \cdots & x_{2k} \\ \vdots & \vdots & \vdots & & \vdots \\ 1 & x_{n1} & x_{n2} & \cdots & x_{nk} \end{pmatrix}$$

and

32.19
$$\mathbf{y} = \begin{pmatrix} y_1 \\ y_2 \\ \vdots \\ y_n \end{pmatrix}, \quad \beta = \begin{pmatrix} \beta_0 \\ \beta_1 \\ \vdots \\ \beta_m \end{pmatrix}$$

The method of least squares. *Multiple regression.*

The coefficient vector $\beta = (\beta_0, \ldots, \beta_k)$ of the hyperplane $y = \beta_0 + \beta_1 x_1 + \cdots + \beta_k x_k$ which best fits the given observations in the sense of minimizing the sum of the square deviations,

$$\left(\mathbf{y} - \mathbf{X}\beta\right)'\left(\mathbf{y} - \mathbf{X}\beta\right)$$

is given by

$$\beta = (\mathbf{X}\mathbf{X}')^{-1}\mathbf{X}'\mathbf{y}$$

### References

See e.g. Judge et al. (1988), or Hogg and Craig (1978).

# Bibliography

Anton, H. : *Elementary Linear Algebra*, 5th ed., Wiley & Sons (1987).

Bartle, R. G. : *The Elements of Real Analysis*, 2nd ed., Wiley & Sons (1976).

Beavis, B. and I. Dobbs: *Optimization and Stability Theory for Economic Analysis*, Cambridge University Press (1990).

Bellman, R. : *Dynamic Programming*, Princeton University Press (1957).

Black, F. and M. Scholes: "The pricing of options and corporate liabilities", *Journal of Political Economy*, May-June, (1973).

Blackorby, C. and R. R. Russell: "Will the real elasticity of substitution please stand up? (A comparison of the Allen/Uzawa and Morishima Elasticities)", *American Economic Review*, Vol. 79, no 4 (1989).

Braun, M. : *Differential Equations and their Applications*, 3rd ed., Springer (1983).

Burmeister, E. and H. R. Dobell: *Mathematical Theories of Economic Growth*, Macmillan (1970).

Christensen, L. R., D. Jorgensen and L. J. Lau: " Transcendental logarithmic utility functions ". *American Economic Review*, Vol. 65, no 3, 367-383, (1975).

Deaton, A and J. Muellbauer: *Economics and Consumer Behaviour*, Cambridge University Press (1980).

Dhrymes, P. J. : *Mathematics for Econometrics*, Springer (1978).

Dickson, L. E. : *Theory of Equations*, Wiley & Sons (1939).

Dixit, A. K. : *Optimization in Economic Theory*, 2nd ed., Oxford University Press (1990).

Edwards, C. H. and D. E. Penney: *Calculus and Analytic Geometry*, Prentice-Hall (1990).

Faddeeva, V. N. : *Computational Methods of Linear Algebra*, Dover Publications, Inc. (1959).

Feichtinger, G. and R. F. Hartl: *Optimale Kontrolle Ökonomischer Prozesse*, Walter de Gruyter (1986).

Fleming, W. H. and R. W. Rishel: *Deterministic and Stochastic Optimal Control*, Springer (1975).

Førsund, F. : "The homothetic production function". *Scandinavian Journal of Economics*, vol. 77, 234-244, (1975).

Friedman, J. W. : *Game Theory with Applications to Economics*, Oxford University Press (1986).

Fuss, M. and D. McFadden (eds.): *Production Economics: A Dual Approach to Theory and Applications*, Vol. I, North-Holland (1978).

Gantmacher, F. R. : *The Theory of Matrices*, Vol. 1, New York, Chelsea Publishing Co. (1959).

Goldberg, S. : *Introduction to Difference Equations*, Wiley & Sons (1961).

Hadar, J. and W. R. Russel: " Rules for ordering uncertain prospects ". *American Economic Review*, Vol. 59, no 1, 25-34, (1969).

Hardy, G. H., J. E. Littlewood, and G. Pólya: *Inequalities*, Cambridge University Press (1952).

Hildebrand, F. B. : *Finite-Difference Equations and Simulations*, Prentice-Hall (1968).

Hildenbrand, W. : Core and Equilibra of a Large Economy, Princeton University Press (1974).

Hildenbrand, W. and A. P. Kirman: *Introduction to Equilibrium Analysis*, North-Holland (1976).

Hogg, R. V. and A. T. Craig: *Introduction to Mathematical Statistics*, 4th ed., Macmillan (1978).

Huang, Chi-fu and R. H. Litzenberger: *Foundations for Financial Economics*, North-Holland (1988).

Intriligator, M. D. : *Mathematical Optimization and Economic Theory*, Prentice-Hall (1971).

Johansen, L. : *Production Functions*, North-Holland (1972).

Judge, G., R. C. Hill, W. Griffiths, H. Lutkepohl, and T-C. Lee : *Introduction to the Theory and Practice of Econometrics*, Wiley & Sons (1988).

Kamien, M. I. and N. I. Schwartz: *Dynamic Optimization. The Calculus of Variations and Optimal Control in Economics and Management*, 2nd ed. North-Holland (1991).

Karatzas, I. and S. E. Shreve: *Brownian Motions and Stochastic Calculus*, Springer (1988).

Kreps, D. M. : *A Course in Microeconomic Theory*, Harvester Wheatsheaf (1990).

Lang, S. : *Linear Algebra*, 3rd ed, Addison-Wesley (1987).

Luenberger, D. G. : *Introduction to Linear and Nonlinear Programming*, Addison Wesley (1973).

Magnus, J. R. and H. Neudecker: *Matrix Differential Calculus with Applications in Statistics and Econometrics*, Wiley & Sons (1988).

McDonald, R. and D. Siegel: "Option pricing when the underlying asset earns a below-equilibrium rate of return: A note." *Journal of Finance*, vol 39, 261-265, (1984).

Nikaido, H. : *Convex Structures and Economic Theory*, Academic Press (1968).

Nikaido, H. : *Introduction to Sets and Mappings in Modern Economics*, North-Holland (1970).

Norstrøm, C. J. : "A sufficient condition for a unique nonnegative internal rate of return." *Journal of Financial and Quantitative Analysis*, 1835-1839, June (1972).

Øksendal, B. : *Stochastic Differential Equations*, 2nd ed., Springer(1989).

Parthasarathy, T. : *On Global Univalence Theorems*. Lecture Notes in Mathematics. No. 977, Springer (1983).

Pontryagin, L. S. : *Ordinary Differential Equations*, Addison-Wesley (1962).

Protter, P. : *Stochastic Integrals and Differential Equations*, Springer (1990).

Rothschild, M. and J. Stiglitz: "Increasing risk: (1) A definition." *Journal of Economic Theory* (1970).

Rudin, W. : *Principles of Mathematical Analysis*, 2nd ed., McGraw-Hill (1982).

Scarf, H. (with the collaboration of T. Hansen): *The Computation of Economic Equilibria*. Cowles Foundation Monograph, 24, Yale University Press (1973).

Seierstad, A. and K. Sydsæter: *Optimal Control Theory with Economic Applications*, North-Holland (1987).

Sharpe, W. F. : "Capital asset prices: A theory of market equilibrium under conditions of risk", *Journal of Finance*, 425-442, Sept. (1964).

Shephard, R. W. : *Cost and Production Functions*, Princeton University Press (1970).

Silberberg, E. : *The Structure of Economics. A Mathematical Analysis*, McGraw-Hill (1978).

Sneddon, I. N. : *Elements of Partial Differential Equations*, McGraw Hill, (1957).

Takayama, A. : *Mathematical Economics*, 2nd ed., Cambridge University Press (1985).

Varaiya, P. P. : *Notes on Optimization*, Van Nostrand Reinhold (1972).

Varian, H. : *Microeconomic Analysis*, Norton (1989).

Wan Jr., H. Y. : *Economic Growth*, Harcourt Brace Jovanovitch (1971).

# Index

absolute risk aversion, 139
active constraints, 73
AIDS (almost ideal demand system), 132
Allen-Uzawa's elasticity of substitution, 19
alpha($\alpha$) level of significance, 152
American call option, 141
angle (between vectors), 91
annuity, 135
arithmetic mean, 27
arithmetic series, 31
Arrow-Pratt risk premium, 139
Arrow's sufficiency condition 79, 81
asymptotes,
    for hyperbolas, 2
    general definition, 3
asymptotically unbiased estimator, 151
autonomous system of differential equations, 51

basis (for a subspace), 90
Bayes rule, 149
Bernoulli's differential equation, 48
Beta distribution, 155
Beta function, 39
bias, 151
binding constraint, 73
binomial distribution, 153
binomial coefficients, 33
binormal distribution, 154
Black and Schole's option pricing model, 141
bordered Hessian, 62, 68

bounded set, 55
Brouwer's fixed point theorem, 24

capital asset pricing model, 141
Cardano's formulas, 1
catching up optimality, 82
Cauchy-Schwarz's inequality, 27, 28
Cauchy's criterion (for convergence), 55
Cauchy sequence, 55
Cayley-Hamilton's theorem, 105
CES-function, 126
chain rule,
    for derivatives, 13
    for elasticities, 18
change of variable (in integrals),
    one variable, 35, 38
    several variables, 40, 41
characteristic polynomial, 103
characteristic root (or eigenvalue), 103
characteristic vector (or eigenvector), 103
Chebyshev's inequality, 27, 151
chi-square distribution, 155
circle, 2
$C^k$-function, 13
closed graph (of a correspondence), 57
closed set, 55
closure, 58
Cobb-Douglas function, 125
cofactor, 99
compact set, 56
comparison test for convergence
    of integrals, 38
    of series, 32

compensating variation, 131
complementary slackness, 74
complex exponential function, 7
complex number, 6
  conjugate, 6
  modulus, 6
  $n$th roots of, 7
  polar form of, 7
  trigonometric form, 7
compound amount of a principal, 135
concave function, 59
conditional density, 151
conditional factor demand functions, 123
conditional probability, 149
cone, 15
conics, 2
consistency (of an estimator), 151
constraint qualification, 73, 74
consumer demand functions, 129
continuity
  one variable, 9
  several variables, 56
contraction mapping, 24
convergence,
  for functions, 9
  for sequences, 55, 56
  for series, 31
convex set, 59
convex function, 59
correlation coefficient, 151
correspondence, 57
cosine function, 4
cost function, 123
cost minimization, 123
cotangent function, 4
counting rule, 22
covariance, 150
Cramer's rule, 96
critical points, 65
critical region (statistical testing), 152
cubic equation, 1
current value Hamiltonian, 80

decomposable matrix, 110
definite integrals, 38, 40
degrees of freedom
  for systems of equations, 22
  in statistics, 155
De Moivre's formula, 7

derivative, 10
derivative (partial), 13
Descartes's rule of signs, 2
determinant, 93, 94
diagonalizable matrix, 104
diagonal matrix, 97
differentiable function
  one variable, 10
  several variables, 14
differential
  one variable, 12
  several variables, 13
dimension (of a subspace), 90
directional derivative, 15
directional elasticity, 18
discount factor, 136
distance,
  in $R^2$, 3
  in $R^n$, 91
dominant diagonal matrix, 112
dot product, 90
double integral, 40
duality theorem (in LP), 72
dual problem (LP), 71
dynamic programming problem, 85

$e$, 4
effective rate of interest, 135
eigenvalue, 103
eigenvector, 103
elasticity of a function, 17
elasticity of substitution
  two variables, 18
  several variables, 19
  Allen-Uzawa's, 19
  Morishima's, 20
ellipse, 2
endogenous variables, 21
envelope theorem, 120, 121
equilibrium point
  for differential equations, 51
  for an economic system, 119
equivalent variation, 131
estimator, 151
Euler's equation, 77, 78
Euler's formulas, 7
Euler's theorem (on homogeneous functions), 15
European call option, 141, 142

## Index

exchangeability property (game theory), 147
existence and uniqueness theorem for differential equations, 53, 54
exogenous variables, 21
expectation, 149, 150
expenditure function, 130
exponential distribution, 154
extreme value theorem (or Weierstrass's theorem), 56, 65

factor demand functions, 124
F-distribution, 155
first degree stochastic dominance, 139
fixed coefficient function, 126
fixed point theorems, 24
Frobenius's inequality, 100
Frobenius root (of a matrix), 110
functional dependence, 22
fundamental theorem of algebra, 2

Gale-Nikaido theorems, 24
Gamma distribution, 155
Gamma function, 39
geometric distribution, 155
geometric mean, 27
geometric series, 31
global inverse function theorem (Hadamard), 23
gradient, 15, 117

Hadamard's theorem, 23
Hadar and Russell's theorem, 140
Hamiltonians,
    continuous control theory, 79
    discrete control theory, 86
Hamilton-Jacobi-Bellman's equation, 143
harmonic mean, 27
Hawkins-Simon conditions, 111
Hessian matrix, 60, 117
Hicksian demand function, 130
Hölder's inequality, 27, 28
homogeneous (differential) equation, 47
homogeneous functions, 15
homothetic function, 15
Hotelling's lemma, 125
hyperbola, 2
hyperbolic sine, 6
hyperbolic cosine, 6

hypergeometric distribution, 153

idempotent matrix, 99, 109
identity matrix, 97
implicit function theorem, 14, 21
indecomposable matrix, 110
indefinite integrals, 35
indefinite quadratic form, 106
indirect utility function, 129
infinite horizon problem
    control theory, 82
    dynamic programming, 85
infinum, 57
inner product, 90
integral test (for series), 31
integration by parts, 35, 38
integration by substitution, 35, 38
interest factor, 136
interior point, 55
intermediate value theorem, 9
internal rate of return, 136
inverse function theorem
    global version, 23
    local version, 23
inverse (of a matrix), 99
invertible (matrix), 99
involutive matrix, 99
Itô's formula, 142, 143

Jacobian determinant, 23
Jacobian matrix, 21, 22
Jensen's inequality, 28
Jordan's decomposition theorem, 105

Kakutani's fixed point theorem, 24
Kronecker product, 113
Kuhn-Tucker's necessary conditions, 74, 76
Kuhn-Tucker's sufficient conditions, 73, 75

Lagrangean problem, 67, 120
Lagrangean function (Lagrangean), 67, 73, 120
Lagrangean multipliers, 69, 73
Law of the minimum, 126
leading principal minors, 96
Legendre's condition, 77
Leibniz's formula, 39

Leontief function, 126
Leontief systems, 111
LES (Linear expenditure system), 132
level curves (and slopes), 14
L'Hôpital's rule, 11
Liapunov function, 52
Liapunov theorems, 52, 53
lim inf, 58
lim sup, 58
linear expenditure system (LES), 132
linear independence
 of functions, 48
 of vectors, 89
linear quadratic control problem, 81
linear system of equations, 25
local inverse function theorem, 23
local saddle point theorem (differential equations), 53
logarithm, 4
logistic equation, 47
lognormal distribution, 154
Lotka-Volterra models, 53
lower hemicontinuity, 57

Maclaurin series, 32
Maclaurin's formula, 32
Mangasarian's sufficiency conditions, 79
marginal rate of product transformation (MRPT), 18
marginal rate of substitution (MRS), 18
marginal rate of technical substitution (MRTS), 18
matrix, 97
 anti-symmetric, 99
 cofactor, 99
 decomposable, 110
 diagonal, 97
 diagonalizable, 104
 dominant diagonal, 112
 idempotent, 99, 109
 identity, 97
 indecomposable, 110
 inverse, 99
 involutive, 99
 Jacobian, 21, 22
 norms of, 101
 orthogonal, 99, 109
 partitioned, 101
 permutation, 110
 rank of, 100
 scalar, 97
 square, 97
 symmetric, 99
 trace of a, 100
 transpose of, 98
 unit, 97
 zero, 98
matrix inversion pairs, 100
matrix norms, 101
maximized Hamiltonian, 79, 81
maximized current value Hamiltonian, 81
maximum principle (continuous time)
 current value formulation, 80
 fixed time interval, 79
 free terminal time, 80, 81
 infinite horizon, 82
 with scrap value, 80
maximum principle (discrete time), 86, 87
mean square error, 151
mean value theorem, 11
 generalized, 11
method of least squares (multiple regression), 156
minimax theorem (game theory), 146
Minkowski's inequality, 28
Minkowski's separation theorem, 59
minors, 95
mixed strategy Nash equilibrium, 145
Morishima's elasticity of substitution, 20
multidimensional normal distribution, 154
multinomial distribution, 153
multiple regression, 156

Nash equilibrium, 145
natural logarithm, 4
$n$-ball, 55
negative (semi-) definite quadratic forms (matrices), 106
Newton's approximation method, 3
Newton's binomial formula, 33
$n$-integral, 40
nonnegative matrix, 110
norm
 of a matrix, 101
 of a vector, 90
normal distribution, 154
normal (to a curve), 4
normal (to a hyperplane), 16

Norstrøm's rule, 137
$n$-person game, 145
$n$th root (of a complex number), 7
null hypothesis, 152

open set, 55
orthogonal matrix, 99, 109
overtaking optimality, 82

parabola, 3
partial derivative, 13
partial differential equation, 54
partial elasticity, 17
partitioned matrix, 101
passus equation, 18
payoff matrix (game theory), 146
periodic functions, 3
permutation matrix, 110
pointwise convergence, 56
Poisson distribution, 154
polynomial, 1
  rational zeros of, 2
positive definite quadratic form subject to linear constraints, 107
positive matrices, 110
positive quasi-definite matrix, 24
positive (semi-) definite quadratic forms (matrices), 106
power of a test, 152
powers, 4
present discounted value (PDV), 136
present value, 135
primal problem (LP), 71
principal minors, 96
probability density function, 149
profit function, 124
projection, 109
projective differential equation, 47
Puu's equation, 125

quadratic form, 105
quadratic equation, 1
quasi-concave programming, 74
quasi-concave function, 61
quasi-convex function, 62

radians, 4
Ramsey's growth model, 137
rank (of a matrix), 100

rational zeros of polynomials, 2
ratio test (for series), 31
recurrence relation, 43
relative risk aversion, 139
Riccati's equation, 48, 81
Rothschild-Stiglitz's theorem, 140
Routh-Hurwitz's stability conditions, 51
  modified conditions, 52
Roy's identity, 129

saddle point
  differential equations, 53
  for $f(x_1, \ldots, x_n)$, 66
  of the Lagrangean, 74
scalar matrix, 97
scalar product, 90
Schur's theorem, 45
Schwartz's inequality, 90
second degree stochastic dominance, 140
separable differential equation, 47
separating hyperplane theorems, 59
series, 31
shadow elasticity of substitution, 19
shadow prices
  in LP, 72
  in static optimization, 69
Shephard's lemma, 123
Simpson's formula, 39
simultaneous diagonalization, 108
sine function, 4
single consumption $\beta$ asset pricing equation, 141
singular value decomposition theorem, 108
Slater condition, 74
Slutsky equation, 131
Solow's growth model, 137
span (of a set of vectors), 90
spectral theorem (for matrices), 104
sporadically catching up, 82
stability
  difference equations, 44, 45
  differential equations, 50, 51
standard deviation, 150
standardized normal distribution, 154
stationary points, 65
statistical testing, 152
Stirling's formula, 39
stochastic control problem, 143

stochastic independence, 149, 151
stochastic integral, 142
Student $t$-distribution, 155
subsequence, 55
subspace, 90
substitution matrix, 124
summation formulas, 33, 34
supply function, 124
supremum, 57
Sylvester's inequality, 100
symmetry of graphs, 3

tangent function, 4
tangent hyperplane, 16
tangent (to a curve), 4
Taylor series, 32, 33
Taylor's formula, 32
trace (of a matrix), 100
transformation, 23, 56
translog cost function, 127
translog indirect utility function, 133
trapezoid formula, 39
triangle inequalities, 27
trigonometric functions, 4
trigonometric formulas, 5, 6
two-person game, 146
type I error, 152
type II error, 152

uniform convergence, 56
uniform rectangular distribution, 155
unit matrix, 97
upper hemicontinuity, 57
upper level set, 61
utility maximization, 129

value function
  control theory, 80
    discrete dynamic programming, 85, 86
    static optimization, 69, 75, 120
Vandermonde's determinant, 94, 95
variance, 150
variation of parameters, 50
vec-operator, 114

Weierstrass's theorem, 56, 65

Young's (or Schwarz's) theorem, 13

zero sum game, 146

# A New Comprehensive Textbook

**H. Lütkepohl,** University of Kiel

# *Introduction to Multiple Time Series Analysis*

1991. XXI, 526 pp. 34 figs. Softcover DM 78,-
ISBN 3-540-53194-7

This graduate level textbook deals with analyzing and forecasting multiple time series. It considers a wide range of multiple time series models and methods. The models include vector autoregressive, vector autoregressive moving average, cointegrated, and periodic processes as well as state space and dynamic simultaneous equations models. Least squares, maximum likelihood, and Bayesian methods are considered for estimating these models. Different procedures for model selection or specification are treated and a range of tests and criteria for evaluating the adequacy of a chosen model are introduced. The choice of point and interval forecasts as well as innovation accounting are presented as tools for structural analysis within the multiple time series context.

This book is accessible to graduate students in business and economics. In addition, multiple time series courses in other fields such as statistics and engineering may be based on this book. Applied researchers involved in analyzing multiple time series may benefit from the book as it provides the background and tools for their task.
It enables the reader to perform his or her analyses in a competent and up-to-date manner. It also bridges the gap to the difficult technical literature on the topic.

☐ Heidelberger Platz 3, W-1000 Berlin 33, F. R. Germany ☐ 175 Fifth Ave., New York, NY 10010, USA
☐ 8 Alexandra Rd., London SW19 7JZ, England ☐ 26, rue des Carmes, F-75005 Paris, France
☐ 37-3, Hongo 3-chome, Bunkyo-ku, Tokyo 113, Japan ☐ Room 701, Mirror Tower, 61 Mody Road, Tsimshatsui, Kowloon, Hong Kong ☐ Avinguda Diagonal, 468-4°C, E-08006 Barcelona, Spain

# Economic Theory

**Managing Editor:** C. D. Aliprantis, Indianapolis

**Co-Editors:** D. Cass, Philadelphia; D. Gale, Boston; M. Majumdar, Ithaca; E. C. Prescott, Minneapolis; N. C. Yannelis, Champaign; Y. Younès, Paris

**Exposita Editor:** M. Ali Khan, Baltimore

**Economic Theory** aims to be the journal at the forefront of modern research in economics. It provides an outlet for research
- in all areas of economics based on rigourous theoretical reasoning, and
- on specific topics in mathematics which is motivated by the analysis of economic problems.

In its first year, **Economic Theory** has been able to publish 26 outstanding research papers. In order to provide you with even more top-quality information, the volume of **Economic Theory** will be increased by 50% in 1992.

Membership to the **Society for the Advancement of Economic Theory** includes subscription to **Economic Theory**.

Membership applications for and further information on the **Society for the Advancement of Economic Theory** may be obtained from: Society for the Advancement of Economic Theory, Department of Economics, University of Illinois, 330 Commerce Building, 1206 South Sixth Street, Champaign, IL 61820, USA

**Papers to Appear in Volume 2:**

**S. Aiyagari, N. Wallace:** Fiat Money in the Kiyotaki-Wright Model.

**C. D. Aliprantis, C. R. Plott:** Competitive Equilibria in overlapping Generations Experiments.

**F. Allen, D. Gale:** Measurement Distortion and Missing Contingencies in Optimal Contracts.

**V. Bala, M. Majumdar:** Chaotic Tatonnement.

**J. Benhabib, R. Radner:** The Joint Exploitation of a Productive Asset: A Game-Theoretic Approach.

**T. Kehoe, D. Levine, P. Romer:** On Characterizing Equilibria of Economies with Externalities and Taxes as Solutions to Optimization Problems.

**M. Peters:** On the Efficiency of Ex Ante and Ex Post Pricing Institutions.

**D. Ray, P. Streufert:** Dynamic Equilibria with Unemployment Due to Undernourishment.

**D. Saari:** The Aggregate Excess Demand Function and Other Aggregation Procedures.

---

☐ Please bill me
☐ Please charge my credit card
　☐ Eurocard/Access/MasterCard
　☐ American Express
　☐ Visa/Barclaycard/BankAmericard
　☐ Diners Club

Number:

| | | | | | | | | | | | | | | |
|--|--|--|--|--|--|--|--|--|--|--|--|--|--|--|

Valid until: _____

Please hand this order form to your bookseller or return to:
Springer-Verlag, Journals Marketing, Tiergartenstr. 17, W-6900 Heidelberg

**Economic Theory** ISSN 0938-2259 Title No. 199

☐ Please enter my subscription beginning with Volume 2 (1992, 4 issues) DM 398,- (suggested list price) plus carriage charges: FRG DM 9,20, other countries DM 20,20

☐ Please send me a free sample copy

Name/Address _____

Date: _____ Signature: _____

Please note: I realize that I may cancel subscriptions within ten days by writing to the address on the order form, whereby the postmark date will suffice as proof that the cancellation was made within the deadline. I confirm my understanding of this with my second signature.

2nd Signature: _____

## Springer-Verlag
Berlin Heidelberg NewYork London Paris Tokyo HongKong Barcelona Budapest